Buxton A Black Utopia in the Heartland

T0346018

A BUR OAK BOOK

AN EXPANDED EDITION

Buxton

A Black Utopia in the Heartland

Dorothy Schwieder,
Joseph Hraba,
and Elmer Schwieder

University of Iowa Press
Iowa City

University of Iowa Press, Iowa City 52242

Printed in the United States of America
Buxton: Work and Racial Equality in a Coal Mining Community
was published in 1987 by Iowa State University Press.
http://www.uiowa.edu/uiowapress

The publication of this book was generously supported by
the University of Iowa Foundation.
Printed on acid-free paper

Library of Congress Cataloging-in-Publication Data
Schwieder, Dorothy, 1933–
Buxton: a Black utopia in the heartland / by Dorothy
Schwieder, Joseph Hraba, and Elmer Schwieder.
Expanded ed.
p. cm.—(A Bur oak book)
Original subtitle: Work and racial equality in a
coal mining community.
Includes bibliographical references (p.) and index.
ISBN 0-87745-852-9 (pbk.)
1. Coal miners—Iowa—Buxton. 2. Buxton (Iowa)—Race
relations. 3. Buxton (Iowa)—Social conditions. I. Hraba,
Joseph. II. Schwieder, Elmer, 1925– III. Title.
IV. Series.
HC8039.M62U66825 2003
305.9′622—dc21 2002041616

03 04 05 06 07 P 5 4 3 2 1

CONTENTS

A BUXTON RETROSPECTIVE
Introduction to the 2003 Edition

IN 1980, when we began to research the history of Buxton, a coal mining community that existed in southeastern Iowa between 1900 and the early 1920s, it quickly became apparent that the Buxton story was unusual in several respects. Buxton had been, after all, the largest unincorporated coal mining community in the state. Most important, however, was that Buxton's population of 5,000 contained a majority of African Americans, a highly unusual racial composition for a state which at the time was about 99 percent white. Although Buxton was essentially abandoned early in the 1920s, even forty years later former residents returned to the town site to hold picnics, greet former neighbors and friends, and reminisce about the prosperous and happy times they had known there. Among former residents, the community became almost legendary. In 1980, when we began researching the town's history, many of the African Americans we spoke to throughout Iowa proclaimed that Buxton had been a utopia for its black citizens.

A central task of our research was to answer a basic, fundamental question: Was Buxton the utopia that former residents proclaimed it to be? Supported by a grant from the Department of the Interior, we began our research in the summer of 1980.[1] A major part of that work was interviewing former Buxton residents, both black and white; in total, we interviewed seventy-five people. We also collected archival data including census records, state mining reports, newspaper articles, manuscripts, and photographs. As we completed the research and analyzed our findings, we felt the evidence did, indeed, support the utopian claims of former residents. The Buxton mines had provided a racially integrated workplace and, in keeping with the United Mine Workers policy, blacks and whites there received equal pay for equal work. Racial equality was also evident in Buxton's residential areas, where blacks and whites lived side by side in similar company housing. Moreover, both blacks and whites operated businesses in Buxton, and black children attended integrated public schools. Given these circumstances, we concluded that Buxton had been a community where racial harmony prevailed and where equal work, equal pay, and equal living conditions existed for African Americans. It seemed to us that the remembrances of former residents rang true: Buxton had been a utopia for black Iowans.

Now, some eighty years after the town's demise and sixteen years after the original publication of our findings, the history of Buxton remains a compelling, intriguing story that continues to capture people's imaginations in Iowa and elsewhere. The Buxton story still produces a sense of awe that a racially integrated community could have existed in Iowa in the early twentieth century. For the people who lived there, particularly African Americans, Buxton seemed a virtual oasis in the midst of an otherwise hostile world. This reprinting of Buxton's history under the new title *Buxton: A Black Utopia in the Heartland* provides us an opportunity to reflect not only on the original study, published in 1987, but perhaps more important, to remember the many former Buxton residents who shared their wonderful memories with us. Most of them are no longer living, but we hope this reprinting of their stories will serve their memories well.

For us, the interviews were the most compelling and significant part of the Buxton data. Although recorded some twenty years ago, the voices of the men and women who once lived in Buxton still resonate with exuberance, affection, and poignancy. Dorothy Collier, who lived in Buxton as a child, later stated that Buxton "was kind of heaven to me, in a way of speaking, 'cause the memories were so nice."[2] Hucey Hart, a miner in both Buxton and nearby Haydock, remembered that in Buxton, with its many businesses and organizations, "you didn't want for nothing." While Dorothy Collier and Hucey Hart were African Americans, Archie Harris was a white man who also perceived Buxton to be an exceptional community. Harris commented that Ben Buxton "must have just loved the colored people . . . because he made everything so nice for them here in Buxton." Former residents related in their own words how Buxton provided not only an accepting workplace for both blacks and whites but also comfortable, integrated housing; good medical care; and steady work with good pay. The interviews also covered the fact that while life in Buxton marked a golden time in the lives of most of its African Americans, life after Buxton—in communities like Cedar Rapids and Waterloo as well as in Milwaukee, Detroit, and Chicago—held little of the acceptance and stability they had known in this black utopia in the heartland.

It was a real privilege to meet and talk with so many former Buxton residents during the 1980s, and revisiting these experiences now brings back warm memories of the people who shared their life stories with us. Although all informants, both black and white, provided valuable information, we detected something of a difference in the two perspectives of life there. For white families, Buxton had been merely a good place to live, with its comfortable housing and modern mining facilities; for black families, Buxton had provided an entirely new experience of racially integrated institutions and other amenities. Particularly for the many black families who had migrated from the South, Buxton seemed almost a perfect place. The result is that while both sets of interviews provided important details of life there, those with African Americans seemed infused with a unique emotional quality and depth of feeling.

We had initially hoped to locate at least a hundred former residents; in retrospect, however, we were probably fortunate to find seventy-five, given that Buxton had essentially ceased to exist nearly sixty years before we began our interviews. The process of obtaining names was varied and sometimes serendipitous. Indeed, the initial contact with former residents of Buxton had been made purely by chance. During the late 1970s, when Dorothy Schwieder was doing research on the topic of Iowa coal mining, a secretary at the State Historical Society building in Des Moines introduced her to three members of the Buxton Iowa Club, a Des Moines group created to preserve the memories of Buxton. Dorothy Collier, Marnetta Skipper, and Eula Miller, three of its members, had come to the state archives looking for material on Buxton's history. A meeting between the four women ended with an invitation for Schwieder to visit the club and to interview club members. Iowa's historic preservation officer, Adrian Anderson, soon initiated a grant proposal for funds to underwrite a study of the Buxton community. Once the Department of the Interior had awarded the grant, the Buxton Iowa Club shared the names of its members and other former Buxton residents. We soon discovered that whenever we did an interview, we came away with additional names of Buxtonites.

Throughout the research phase, people were extremely generous with their time and also with their photographs of Buxton. In Waterloo, we conducted two interviews with Vaeletta London Fields, learning much about her life as well as that of her mother, Minnie B. London, a well-known Buxton school teacher. Vaeletta, a petite, energetic woman in her early eighties, fairly bubbled with excitement when she talked about her life in Buxton. Through the years, she had put together many scrapbooks of pictures and newspaper clippings related to her life in Waterloo in addition to material on Buxton. It seemed that whatever topic we discussed, Vaeletta had a scrapbook item covering the event. Vaeletta's material was especially valuable in helping us understand her life and the life of other former Buxton residents who had moved to Waterloo, particularly in the area of race relations. For Vaeletta, life in Waterloo had been vastly different from life in Buxton.

Sometimes we also learned about relatives who lived outside of Iowa. In the course of her interview, Vaeletta mentioned her cousin, Herman Brooks, who had also lived in Buxton but later moved to Chicago. Herman, she announced, would be able to tell us many things, especially about the renowned Buxton ball club, the Buxton Wonders. In animated, almost schoolmarm tones, Vaeletta insisted we *must* go to Chicago and interview Herman. It was wonderful advice. We went there in early July, 1981, when the weather was terribly hot and humid. Knowing that Herman was ninety-seven years old, we wondered how much he would remember of his halcyon days in Buxton. We had no cause for concern; he was in marvelous health and remembered remarkably detailed information. Herman was, in fact, the only interviewee who clearly remembered Muchakinock, a camp operated by Consolidation Coal Company from the 1880s to 1900. Sixteen years old in 1900, Herman was still able to describe for us the moving process from Muchakinock to Buxton and the manner in which housing was assigned to mining families. Just as Vaeletta had predicted, Herman had many memories of the Buxton Wonders, including his own career as a team pitcher. We arrived at Herman's apartment just after lunch and didn't leave until late afternoon.

During the early 1980s, newspapers around the state occasionally ran stories on the Buxton project. These articles led us to additional informants. Ada Baysoar Morgan had lived there as a teenager when her father was superintendent of the Buxton mines. Although she had never been back to Buxton after leaving there around 1914, she had subscribed to the Albia newspaper (Albia was some eight miles from Buxton) even into her nineties. When that paper carried a feature story on the Buxton project, Ada wrote to say she had pictures of the superintendent's home there. We visited her in Peru, Illinois, finding her to be a gracious lady, most hospitable in her manner and pleasant in demeanor. Yet even in her nineties, she displayed a sense of reserve that perhaps reflected her family's social position in Buxton. The interview yielded not only specific details of the Baysoar family's life there, but the recognition that—like all communities—Buxton had a social hierarchy. Ada's family had interacted with only a few other families, all of whom belonged to the company's management level. During her four years in Buxton, she had never been to the company store; moreover, even by the time we interviewed her, she didn't know that Buxton was renowned for its racially integrated practices. She and her family apparently lived in their own insular world, quite separate from the bustling life of the nearby coal camp. As we left Peru, Illinois, that day, we realized we had heard the echoes of a very different life in Buxton than what we had experienced through the stories of most former Buxtonites.

Following Buxton's abandonment, a few residents remained in the area. These people, like interviewees Archie Harris, Alex and Agnes Erickson, and Reuben Gaines, Jr., had not only watched Buxton develop, but had also observed its dismantling process. Their remembrances provided an excellent longitudinal view of Buxton as well as descriptions of Consolidation's later camps of Consol and Haydock. Their oral histories allowed us to tap into a treasure trove of memories covering a wide range of experiences and perspectives before, during, and after life in Buxton.

Perhaps the two people most knowledgeable about the total Buxton experience were Alex and Agnes Erickson. Brother and sister, they had grown up in a coal mining household in East Swedetown, a suburb of Buxton. Alex and his father had worked as miners for Consolidation and Agnes had worked in the company store. Although both Alex and Agnes were in their eighties at the time we interviewed them, they had a remarkable recall of activities at the company store, life in East Swedetown, and the actual work of coal mining itself. Alex especially enjoyed sharing his knowledge of Buxton; he also proved an excellent tour guide as he showed us all the locations for Muchakinock, Buxton, and Haydock, as well as sites for the outlying mines.

The only Buxton businessman we interviewed was Reuben Gaines, Jr., who had owned businesses in both Buxton and Haydock and had also lived in Muchakinock. Although somewhat restricted in his mobility, Reuben insisted on playing several lively ragtime melodies for us on the piano. Reuben also had many pictures of Buxton which he graciously shared with us. Because of his business connections, he had known hundreds of people in the area.

While life in Buxton held myriad opportunities for African Americans as a group, there was also the constant reminder that mining was dangerous work. Mattie Murray's parents had migrated from Virginia to

Muchakinock, and later moved to Buxton. When Mattie's father, Tandy Murray, died, her mother was left to support the fourteen children she still had at home. Mattie remembered the struggles and the hardships her mother faced in providing for her large family. Odessa Booker also remembered difficulties in Buxton after her father was killed in a mining accident there. Odessa's plans to attend high school ended the day of her father's death. She and her younger brother both went to work for the company to help the family cope financially.

In the course of our interviews, we also had the opportunity to see artifacts from the Buxton period. In the early 1980s, we spent an afternoon with Dorothy Collier. As we sat around her dining room table and drank coffee, she brought out a lovely china soup tureen that her mother had used in their family home. For Dorothy, it was a cherished item that provided a direct link back to the family's days in Buxton. For us, it was a concrete reminder that most African American families in Buxton had lived well at a time when blacks in many other places were experiencing poverty and discrimination.

Of all the seventy-five people interviewed, only one person had less than glowing memories of Consolidation. Robert Wheels, known to everyone as "Preach," dismissed any altruism on the part of the company. Wheels believed that because Consolidaton had developed the town as a commercial venture, its only concern was making a profit. In his quiet, gentle way, Wheels allowed that black families in Buxton lived well, but he refused to accept that it was really a utopia. Although he admitted he had a good life there, "still you was a slave in a way of speaking." Wheels added that he was glad he was able to leave Buxton because "if I hadn't left . . . I wouldn't have known how to do nothing but coal mine, or get out and work a little bit on a farm or something like that. The work here [working for the Rock Island Railroad in Des Moines] was an education to me." Wheels, more than any other interviewee, reminded us that Buxton was, indeed, a company town.

In reflecting on the subjects covered in the original study published in 1987, we feel the need to mention at least one topic we omitted: the issue of social class. Interviews and other data yielded little direct information on the subject, yet given what we do know about Buxton society, it is possible to make informed inferences about this social phenomenon. Historical studies show that people in all societies have been differentiated to some degree by characteristics of education, income, occupation, and family background. Sometimes, these differences in life-style which reflect issues of social class have been blatant; at other times they were more subtle; but always, in any society or community, social stratas existed. Certainly social classes existed in Buxton. Ada Morgan's account of life there clearly underscores the point that her life at the top was decidedly different than how she would have lived in the main residential area of the town. Management officials, including R. R. MacRae, Dr. Henderson, and W. H. Wells, along with their families, no doubt interacted socially on many occasions. MacRae, Henderson, and Wells were related by marriage to the Buxton family, indicating that kinship ties also played an important part in determining class structure.

Among African Americans in Buxton, education and occupation seem to have been major considerations in determining class structure. Dr. E. A. Carter, a medical doctor, held a position of great respect within the community. Dr. Carter's wife, Rosa, belonged to the Silver Leaf Club, described by her daughter, Marian, as "one of the most elite black women's groups in Buxton." Marian also remembered that her mother had household help, probably the wife or daughter of a black Buxton miner. George Woodson, an attorney, was also a prominent person in the camp, along with pharmacist B. F. Cooper. Woodson, Cooper, and W. H. and Minnie London were frequently mentioned in the columns of the *Bystander*, which reported regularly on social activities in Buxton. W. H. was a local businessman and Minnie was a teacher and later principal in the Buxton school system. The Londons apparently felt strongly about education, as both of their children graduated from the University of Iowa. While class lines certainly existed, there was considerable social fluidity even though it seems doubtful that social interactions took place between mining families such as the Harts or the Ericksons and members of the black professional class such as Woodson or the Carters.

One variable that undoubtedly affected social interactions was the matter of family size. For women like Robert Wheels's mother, who had fifteen children, life probably centered around home and church. Handling the domestic tasks for fifteen people (two children had died young), in a home without electricity or running water, would have been demanding. Other Buxtonites also had large families. Mattie Murray's mother had seventeen children, while Oliver Burkett's mother had eleven. A number of families had nine or ten children. Some miners' wives contributed to their families' income by earning money through sewing or selling garden produce. Some women took in washing. Lewis Reasby ran a lunch wagon on Buxton's main street and apparently did a thriving business, but the food prepared for sale there was mostly prepared at the Reasby home by his wife and daughters. A relative of Reasby's estimated that on some weekends as many as 200 to 250 chickens were sold, chickens presumably prepared by the Reasby women.

Another point we might well have emphasized more strongly in the 1987 study is the extraordinary collection of black professionals in Buxton, particularly during the town's first decade. In addition to Dr. E. A. Carter, George Woodson, B. F. Cooper, and Minnie B. London, the town also included a dentist, Dr. Linford Willis, and several teachers, including S. Joe Brown and his sister, Sue Brown. Other professional people included the director of the YMCA and several African American ministers. Buxton also included a significant number of black businessmen; some, like George Neal, worked both as a miner and a tailor; several other African Americans published newspapers, again in addition to their work in the mines. It seems that in the first decade of the twentieth century, Buxton acted as a magnet to African Americans, offering exceptional professional and business opportunities as well as employment opportunities for the miners themselves.[3]

While we were able to locate an extensive amount of information about Buxton and its citizens, some aspects of the community remain a mystery. Why, for example, did the community begin to decline around 1914? During World War I, the Iowa coal industry as a whole experienced some years when production was down, but in general, the industry did fairly well. In fact, 1917 represented its peak production year. By 1919, there were signs of trouble in the industry, yet even eight years later, in 1925, Iowa still had 354 underground mines with a total of 11,241 miners. For some reason, the expansion and prosperity that came with World War I was not evident in Buxton or in the other communities that Consolidation created later in Iowa. When, in 1927, Consolidation closed all of its Iowa operations, the reasons were apparent. First, the company had experienced two major strikes, and second, by the latter 1920s, the entire Iowa coal industry was in serious decline.[4]

It is clear that Consolidation made major management changes in 1913 and 1914, and these could have affected Buxton's prosperity. Former residents recalled that after 1914, the company hired more white miners than black miners, and they also hired fewer blacks in positions such as postmaster. The reasons for these changes are not clear. Presumably company policies established in 1900 had been working well, as the company had seemed satisfied with the large number of African American miners before 1914. Ben Buxton's resignation as mine superintendent in 1909 did not seem to bring as much change as did Superintendent E. M. Baysoar's departure in 1913. Whatever the reasons for the policy changes, according to former residents, after 1914 Buxton seemed to offer fewer opportunities for blacks.[5]

Perhaps the greatest mystery surrounds the town's namesake, Ben Buxton. Our efforts to locate background information on him or to discern his possible motivation for creating a predominately black community were largely unsuccessful. We know the basic facts—where he was born, his family background, and when he arrived in Iowa—but beyond that, there is little information. Did Ben Buxton feel something of a mission to help create a racially harmonious community that might act as a model for other businesses? If so, what had influenced his thinking to the point that he deviated so greatly from the norm in Iowa in 1900? Or, was he influenced by the prevailing concepts of the day such as welfare capitalism, which indicated that if you provide comfortable, well maintained housing and good recreation for your workers, these conditions pay dividends in workers that are then more productive? As Robert Wheels believed, was Ben Buxton simply setting up a community that would be financially profitable?

This Buxton retrospective also allows for a comment on the value of doing oral history. For many years, American historians have been skeptical of information obtained through interviews, the general belief being that historical information should be gathered from printed sources rather than spoken words. Fortunately, during the past thirty years that view has been fading as projects such as the Buxton study have revealed the value and uniqueness of material gathered through oral history. While printed sources such as census records, newspapers, and government reports provide invaluable material for any researcher, these sources typically do not include the more personal information needed to complete, and often interpret, the whole story. Census records, for example, show us *when* people lived in Buxton, but they provide no information on the reasons why people moved there, or how they lived once they arrived. The day-to-day routines of men, women, and children—including work, social events, religious affairs, and family and household activities—often become a part of the historical record through information obtained from oral histories. What follows in the chapters on Buxton could be told only because of the collection and analysis of oral history material.

This reprinting of the Buxton book also relates to another aspect of African American history: the availability of historical sources. When we started our research in 1980, there were few published sources on African American history in Iowa. Recently that void has been greatly reduced with the publication of *Outside In: African-American History in Iowa, 1838–2000.* The book, some five hundred pages in length, contains articles on almost every phase of African American history in the state, from black farmers in the early days of statehood, to black contributions in the fields of medicine and law in the twentieth century. While *Outside In* is particularly meaningful to African Americans in Iowa and elsewhere, its publication means that everyone can now read about the many contributions of blacks to all phases of society. Along with *Outside In,* the reprinting of the Buxton book (which has been out of print for over seven years) will allow future generations of Iowans and other Americans to understand that African Americans in this state have a proud history of accomplishments.

Finally, mention should be made of another legacy of the Buxton project. During the fifteen years since *Buxton: Work and Racial Equality in a Coal Mining Community* was first published, we have received dozens of letters, phone calls, and in more recent years e-mails in which people have inquired about family members who once lived in Buxton and, in some cases, have supplied additional information about family members who lived there. A recent response provided information on the Jeffers brothers, two African Americans who ran a restaurant in Buxton. Unlike many Buxton residents, the Jeffers brothers did not come directly from the South, but had lived in Ohio and Indiana before arriving in Iowa. One family member, Andy Jeffers, became a class leader at the St. John's African American Church in Buxton. Another letter told of Swedish American activities in the two Swedetowns and the importance of the Ebenezer Lutheran Church to the Swedish Americans there. Many letters reiterated how Buxton had lived in people's memories as an extraordinary place to call home in the early twentieth century.

We hope this reprinting of the Buxton history will continue to speak to people who had family members there, but we also hope it will reach people who have not yet heard about this special place. We believe the history tells a significant story about people's ability and willingness to live and work together, to share community activities, and, in general, for a pluralistic society to be mostly free of strife and conflict. It seems fitting here to repeat former resident Marjorie Brown's observation: "Buxton was something else. You can imagine how we grieved for it."

NOTES

1. The federal grant was awarded by the Heritage Conservation and Recreation Service of the Department of the Interior. In addition to funding archival research, the grant also funded archaeological research carried out by David M. Gradwohl and Nancy M. Osborn (now Johnsen) of the Department of Sociology and Anthropology at Iowa State University. That research was published in *Exploring Buried Buxton: Archaeology of an Abandoned Iowa Coal Mining Town with a Large Black Population* (Ames: Iowa State University, 1984; rpt. 1990).

2. David Mayer Gradwohl and Nancy Osborn Johnsen, "'A Kind of Heaven to Me': The Neal Family's Experience in Buxton, Iowa," in *Outside In: African-American History in Iowa, 1838–2000*, ed. Bill Silag (Des Moines: State Historical Society of Iowa, 2001), p. 4.

3. Dorothy Schwieder, Joseph Hraba, and Elmer Schwieder, *Buxton: Work and Racial Equality in a Coal Mining Community* (Ames: Iowa State University Press, 1987), pp. 27–28, 32, 80, 96, 100, 109, 207–208.

4. Dorothy Schwieder, *Black Diamonds: Life and Work in Iowa's Coal Mining Communities, 1895–1925* (Ames: Iowa State University Press, 1983), pp. xi, 212.

5. Schwieder, *Black Diamonds,* chapter 7.

ACKNOWLEDGMENTS

We are grateful to many people for assistance with the Buxton project. First, we would like to acknowledge and thank the Heritage Conservation and Recreation Service of the Department of the Interior for a grant for the period June 1, 1980, to May 31, 1982, which underwrote the cost of the interdisciplinary research for this study. The original research personnel included David Gradwohl and Nancy Osborn of the Department of Sociology and Anthropology at Iowa State University, who carried out the archeological portion of the grant. Dorothy Schwieder of the Department of History at Iowa State University, Joseph Hraba of the Department of Sociology and Anthropology, and Elmer Schwieder of the Department of Family Environment conducted the oral histories and collected the archival data.

We are also indebted to many people for the successful completion of the grant and for assistance in preparing the manuscript. We particularly want to thank Adrian Anderson, former Director of the Iowa State Historical Department, for assistance in writing and administering the grant. The Buxton grant proposal was drawn up jointly by the Iowa State University PREPS Office, project personnel, and Adrian Anderson. We also want to thank Iowa State University for contributions to the project in the form of faculty salaries and the use of ISU facilities. Our thanks also go to Lowell Soike and Jack Lufkin of the Iowa State Historical Department, who expedited many necessary grant procedures and answered innumerable questions about grant requirements. Many other individuals helped collect and process the data. Mornetta Skipper, Mary Jane Graves, Joyce Bruce, Pam Bruce, and Donald Graves, Jr., all spent many hours collecting census data. Louise Fedricks and Debbie Henderson, graduate students in the Department of Family Environment, helped process the data and type the oral history transcripts. Jeri Cole and Phyllis Gray, graduate students in the Department of Sociology and Anthropology, also assisted with this work, and we wish to thank the Department of Sociology and Anthropology for the financial support it gave them. We also want to express our gratitude to Audrey Burton, manuscript typist for the Department of History, for patiently typing several drafts of the manuscript. Finally, a special thanks to Linda Kerber, University of Iowa, for her careful reading of the manuscript.

Most of all, we are deeply indebted to all the former Buxton residents who consented to be interviewed. Many former residents allowed us to copy their personal photographs of Buxton and of family members who lived there. We also wish to thank the Buxton, Iowa Club, a group of former Buxton residents who live in Des Moines and assisted with the grant. In particular, we wish to thank Buxton Club member Dorothy Collier, who provided the project team with continual support. Mrs. Collier supplied names of people to be interviewed and set up and took part in many interviews.

Buxton
A Black Utopia in the Heartland

INTRODUCTION

IN 1881 the Chicago and North Western Railroad purchased the Consolidation Coal Company in Muchakinock, Iowa. This move marked the beginning of an association between the railroad and the Iowa coal industry that lasted almost half a century. Through its subsidiary, the Consolidation Coal Company, the Chicago and North Western dominated that industry, controlling the state's three largest coal communities. The first, Muchakinock, existed from the late 1860s to 1900. In 1900 the Consolidation Coal Company abandoned Muchakinock and transferred its operations to the newly developed coal camp of Buxton. Twenty-two years later, Consolidation again made a major move, shifting its operations from Buxton to the new coal camp of Haydock.

While both Muchakinock and Haydock served as important mining centers, Buxton was the largest of the three and exhibited the most unusual social and economic characteristics. Buxton was both a planned community and a model community. It contained a wide variety of ethnic groups, but during the community's early existence blacks predominated. With a large black labor force, Consolidation officials developed an integrated community where blacks and whites lived and worked side by side with little racial strife. Black former residents refer to Buxton as "the black man's utopia in Iowa."

Given the period in which Buxton existed—the first two decades of the twentieth century—the Buxton experience was highly unusual for both Iowa and the nation. At a time when southern blacks were oppressed by Jim Crow laws and grandfather clauses and northern blacks lived in large urban ghettos, blacks in Buxton enjoyed steady employment, above-average wages, good housing, and a minimum, if not a complete absence, of racial discrimination. Because of these highly unusual social and economic conditions, the Buxton experience has both regional and national significance.

At the time of Buxton's inception, coal mining in Iowa was well established. The state ranked fifteenth in the nation in coal production. Iowa's coal industry was closely tied to the state's railroad system; as railroad mileage increased, the number of coal mines also increased. Railroads were the largest consumers of Iowa coal and often developed their own mines, commonly known as captive mines. These mines formed the largest coal operations in the state. Iowa also contained some three hundred small mines, which produced coal for local consumption. The Iowa coal industry prospered from the mid-1880s to the mid-1920s, the time span that included the rise and decline of Buxton.

The most visible evidence of Iowa's coal industry was the presence of dozens of mining camps scattered throughout central and southern Iowa. Before 1900, Iowa's camps typically contained native-born people and immigrants from the British Isles and Sweden. After 1900 the camps' ethnic composition began to change as more and more families immigrated to Iowa from eastern and southern Europe. By 1920 most Iowa coal camps contained a wide diversity of ethnic groups, but Italian- and Croatian-Americans dominated. At the same time, the majority of Iowa camps included at least a small number of blacks. Before 1900 southern blacks who migrated to Iowa had found employment as steamboat and railroad workers as well as coal miners. After 1900 most blacks worked as coal miners, although they remained a minority in the Iowa coalfields. The coal camp of Buxton proved an exception. Five years after its founding, blacks made up 55 percent of Buxton's population. The remainder included many foreign-born people from all parts of Europe. Buxton also contained substantial numbers of native-born whites, many from eastern coal-producing states, to which their parents and grandparents had earlier immigrated from the British Isles.

Another feature that set Buxton apart from other coal camps was the manner in which it was developed. Most coal camps were laid out in a haphazard fashion close to the mine itself. Owners typically established only one mine at a time and housed their work force in a camp located adjacent to the mine. Mining families lived in the very shadow of the coal tipple, exposed to the daily noise and grime of the mining operations. Since Consolidation operated three mines at the time it founded Buxton—and anticipated opening several more—

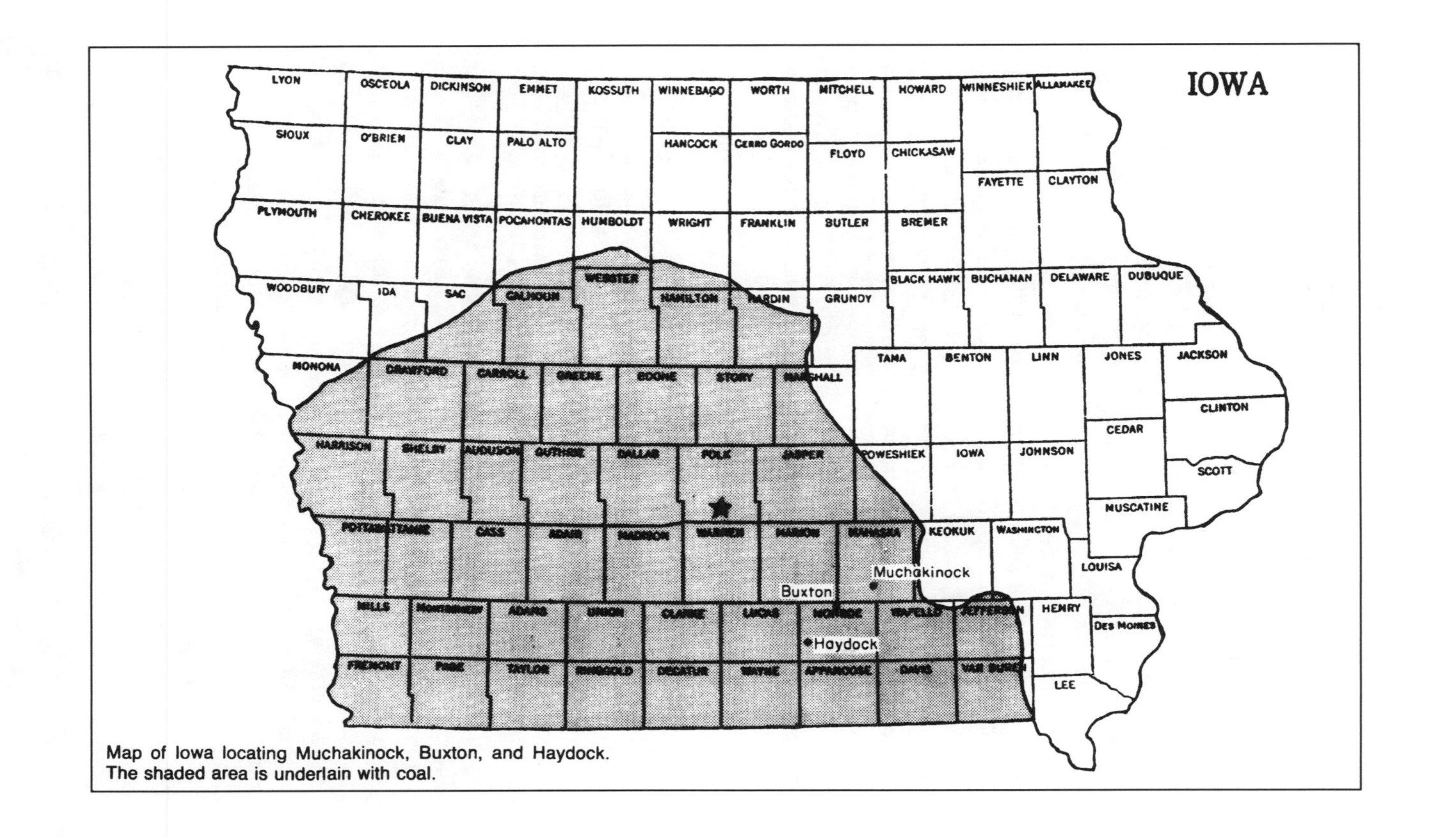

Map of Iowa locating Muchakinock, Buxton, and Haydock.
The shaded area is underlain with coal.

there was no advantage in placing the new community close to any existing mine. Consolidation officials therefore had far more opportunity to select a suitable site for Buxton, taking into account such factors as good drainage, adequate water supply, and the potential for future expansion. Buxton's general layout differed little from that of non–coal mining midwestern communities of a similar size.

Buxton also differed in other ways from surrounding coal camps. Since most coal mines were limited in size, often to 100 or 125 employees, the camps that housed these workers remained small. Along with the miners' homes, the typical Iowa coal camp contained a company store, a pool hall, a miners' union hall, and an elementary school. These coal camps often had little effect on neighboring areas, either economically or socially. Buxton's size, however, and the numerous businesses located there meant that the community had an immediate impact on surrounding areas. Local farmers bought and sold at Buxton's company store, and people living nearby came to Buxton for social events and to attend church. At all times, economic and social exchanges took place between Buxton residents and those in adjacent areas.[1]

Buxton's most unusual characteristic, however, was its high degree of racial harmony. Although towns and cities elsewhere had a similar racial and ethnic mix, few towns exhibited the same type of race relations. In communities with large black populations, most blacks and whites led separate lives, socially and economically. Even then, overt racial antagonism often occurred. This was not the case in Buxton. There blacks and whites worked side by side in local coal mines and lived side by side in company housing with little racial strife. Local public institutions, including schools, were integrated. Former Buxton residents, both black and white, have stated that the Consolidation Coal Company showed no favoritism to white workers in the areas of jobs, wages, or housing. Buxton also contained many black professional and business people, who occupied important economic and social positions. In short, there was little racial stratification, segregation, or strife in Buxton. Because of the general racial situation of the time in which the community existed, the Buxton experience can be understood fully only if placed in a national perspective.[2]

The situation that blacks found in Buxton was in sharp contrast to what they had known in the South. After the Civil War, a racial caste system developed that excluded southern blacks from nearly all occupations except farm labor and domestic service. As a result, many

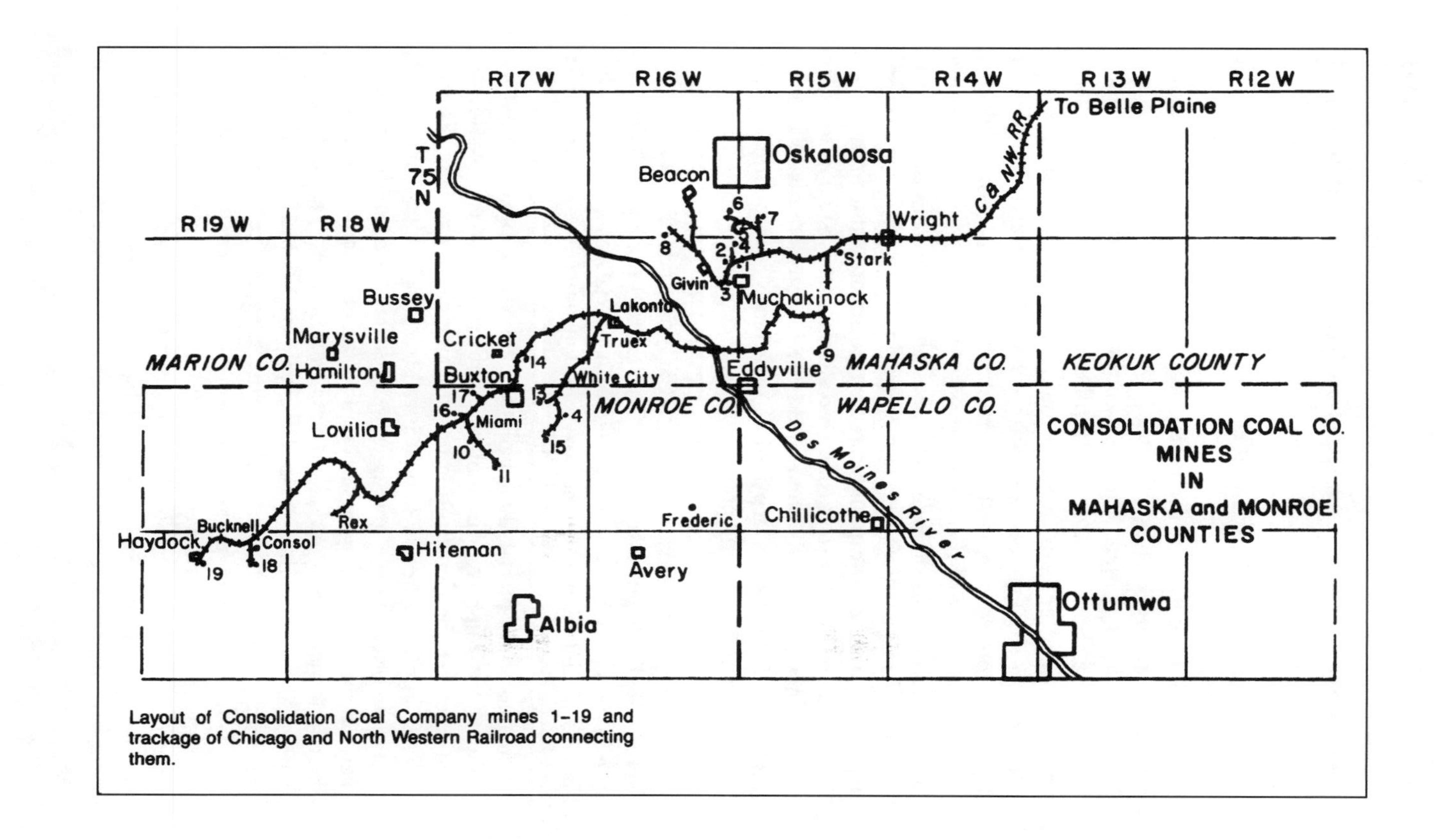

Layout of Consolidation Coal Company mines 1–19 and trackage of Chicago and North Western Railroad connecting them.

black men became tenant farmers or sharecroppers. According to Fred A. Shannon in *The Farmer's Last Frontier*, sharecropping provided black men and their families with barely enough income to survive and no opportunity to advance themselves. Given the particular credit system used in the South—the crop-lien system—many southern farmers not only lived at the lowest poverty level but found it nearly impossible to break out of the poverty cycle. Because most farmers had no cash, they were forced to borrow money to maintain their farming operations. Under the crop-lien system, country merchants annually issued enough credit to southern farmers to purchase supplies for the coming year, accepting as collateral a certain portion of the year's crop. After harvest, farmers often discovered that their debts exceeded the value of their crops; they then had no choice but to carry over the credit into the following year. This situation was often perpetuated for years. Moreover, farmers paid inflated prices at the country stores, further compounding their economic difficulties.[3]

Southern blacks also encountered a racial caste system in southern politics. After Reconstruction ended in 1877, southern blacks increasingly found themselves deprived of their newly won political rights. Georgia in 1877 levied a poll tax of between one and two dollars on eligible voters. Often poor southern blacks found this tax prohibitive. In 1878, for another example, South Carolina established a restriction known as the "eight box system." This called for election officials to use eight boxes to collect ballots, with officials counting only the ballots placed in the correct box. Election officials instructed white voters where to place their ballots but did not instruct blacks, who then frequently placed their ballots in the wrong box. In 1890 the state of Mississippi required all voters to prove that they could read and interpret the Constitution before being eligible to vote. Election officials applied different standards to blacks and whites, and in some instances black college graduates were disfranchised under the rule. Later in the decade, southerners applied the grandfather clause, which in effect stated that one could vote only if one's grandfather had voted. These and other tactics effectively barred blacks from voting throughout most of the South by the turn of the century.[4]

During the same period, southern blacks experienced the advent of Jim Crow laws. Between 1890 and 1920, state after state in the South passed laws that effectively separated blacks and whites in most areas of public life. State laws and local ordinances instructed black people to sit in the rear of busses, to drink from separate water fountains, to eat in separate restaurants, and to attend separate schools.[5] Physical violence against southern blacks also became more frequent, particularly lynching. Between 1880 and 1918, over 2,400 blacks were lynched in southern states. Race riots also occurred. In 1906, for example, a mob attacked blacks in Atlanta, resulting in the death of ten blacks and two whites. For southern blacks, the late nineteenth and early twentieth century was a time of political disfranchisement, very limited economic opportunity, and brutal physical intimidation.[6]

While these conditions prevailed for southern blacks, conditions for northern blacks were sometimes little better. In 1910 nearly 90 percent of American blacks lived in the South, but that year marked the beginning of a massive black migration to the North. Two sociologists, Leonard Broom and Norval Glenn, have described the exodus:

> During the decade following 1910, the great migration of Negroes from the rural South began with the push of a depression in Southern agriculture and the pull of new opportunities for industrial employment in the North. . . . World War I reduced immigration [of Europeans] and increased the demand for new workers, thus creating a labor vacuum into which thousands of Southern workers could move.[7]

Moving into northern cities, black families often found it difficult to obtain employment and housing. Many northeastern cities soon developed black ghettos, where blacks felt trapped and had little hope of escape. Many black families soon discovered that, to survive, wives had to work outside the home, but even then black families often lived in desperate poverty. Elizabeth Pleck discovered that black married women experienced a rate of employment from four to fifteen times higher than immigrant wives. This was due to low income and unemployment among black males. When black men did find work, it was as unskilled laborers or as servants, waiters, or elevator operators, all relatively low-paying jobs. Most black women worked in laundries or as domestics; many took in boarders.[8]

Blacks also went to work in the industrial plants of the North. In 1914, Henry Ford began to hire blacks in his automobile plants and advertised that no worker would be paid less than five dollars per day. Moreover, blacks went to work in steel mills, packing plants, and coal mines. In the latter industry, blacks were used as strikebreakers in many states throughout the Middle West and Northeast.[9]

While the bulk of the black migration from the South took place after 1900, some blacks had already left the South by then, relocating in small communities in the West. In the years just after the Civil War, blacks settled approximately a half-dozen communities in Kansas. These communities were basically agricultural, with black residents purchasing land or taking up government land. According to Robert Athearn, this movement was an orderly one, and after arriving in Kansas participants had some money to establish themselves. But during 1879 and 1880 a second wave of black settlers known as the Exodusters, moved into Kansas. These people were, by contrast, often penniless, and they relied heavily on support from outside groups. Their experience was less successful, and eventually some of them returned to the South or moved on to northern cities. Robert Athearn estimated that the Exodusters numbered around 20,000 people.[10]

For blacks who settled in Buxton, Iowa, however, the experience was quite different. Although the blacks who came to Buxton were, no doubt, predisposed to leave the South, migration was sponsored and planned by the Consolidation Coal Company. Once in Iowa, jobs and homes awaited them. They did not face the long struggle to find employment and housing that many blacks who migrated to the Northeast encountered, and once located in Buxton, they did not experience racial discrimination. This is, then, the story of Iowa's most unusual coal mining community. It was in many ways extraordinary and its impact reached far beyond the borders of the Hawkeye State.

The present study has been an interdisciplinary project from the beginning, involving an American historian, a family sociologist, and a race relations sociologist. Although the present work is primarily a historical study, we have approached the subject from the standpoint of both history and sociology. Our intent is to examine the Buxton experience from numerous perspectives, utilizing not only the traditional historical approach of descriptive narrative and analysis, but also the sociological analysis of family, race relations, and ethnicity. The result is that we have examined the Buxton community both from afar and at close range. Moreover, our interests have led us to examine the population closely, looking specifically at many individuals, and to examine local institutions.

The history of Buxton is actually the history of a series of communities. The first, Muchakinock, preceded Buxton by over thirty years. Policies and procedures established there by the Consolidation Coal Company carried over to Buxton along with most of the camp's population. An understanding of the origin of these policies and the role played by Consolidation is vital to an understanding of the Buxton experience. The Muchakinock experience will be presented in Chapter 1. After 1922 Consolidation shifted its operation from Buxton to yet another coal camp, Haydock, and that community continued until 1928. Many Buxton residents remained with Consolidation and relocated at Haydock. That phase of the Buxton story will also be told. The focus of the study, however, is on the social and economic history of Buxton, with emphasis upon the people who lived there, including their family life, economic life, leisure-time activities, religious activities, class structure, occupational structure, and ethnic composition. The study also includes a physical description of the community and a description and analysis of the role played by the Consolidation Coal Company.

The history of Buxton is in many ways the history of a black community, and numerous historical sources reflect that fact. The most helpful source on blacks in Buxton is the *Iowa State Bystander*, a newspaper published in Des Moines for the state's black community. The *Bystander* provided detailed accounts of activities in Buxton from its founding to around 1920. These included physical decriptions of buildings, feature stories on prominent blacks, and discussions of business and social life. The *Bystander* viewed Buxton as primarily a black community and considered it an ideal place for blacks to live. Unfortunately, there is no comparable record of Buxton's white residents, either native- or foreign-born. The abundance of information on blacks presented the temptation to concentrate on that population to the exclusion of other ethnic groups. We determined, however, that this would not provide the best understanding of the community. Although Buxton was perceived by many people as a predominantly black community, it is particularly important to recognize Buxton's other native-born and foreign-born populations, since the harmonious interaction between blacks and whites is one of the most significant facets of the Buxton story.

The sources used in this study include census data from both the state and federal censuses of Iowa. We used the state censuses of 1895 for Muchakinock and 1905 and 1915 for Buxton, as well as the federal census of 1900 for Muchakinock and 1910 for Buxton. While the census data provide exact information on each Buxton resident—including age, occupation, place of birth, religion, and, in 1915, income and unemployment—these records are not without flaws. It was with great frustration that we discovered that Buxton had not yet come into existence by June 1900 when the federal census was taken in Monroe and Mahaska counties, but rather was organized several months later. Because of this, and because most Muchakinock residents moved to Buxton immediately after its founding, we utilized the 1900 federal census for Muchakinock, presuming the Muchakinock's population would be fairly similar to Buxton's only a few months later. Another drawback to the use of census data was that parts of it have been destroyed. A portion of the 1905 Iowa state census was rendered unusable in a flood, so the Buxton records for that year are only about three-fourths complete. Even with these drawbacks, however, the census records provide valuable insights into Buxton's economic, social, and ethnic composition.

Much information for the study came from seventy-five interviews with former Buxton residents. We had initially hoped to locate a greater number of former residents but, with limited time and money, we were unable to do so. We also discovered that, while there are many persons living who once resided in Buxton, many of these people lived there only during their early childhood and remember little of that experience. We had also hoped to secure a cross section of Buxton's social and economic life through the persons interviewed. We soon discovered, however, that we could not locate a sufficient number of well-informed former residents to achieve that goal. Nevertheless, the seventy-five informants provided an abundance of detail on their lives in Buxton and, in our view, contributed the greatest understanding of Buxton's population.

The combination of census data and interview material has been particularly helpful in understanding Buxton's social and economic composition. Information given in interviews could often be checked against census data for validation. At the same time, information gained from the censuses provided a better sense of questions to be asked in the interviews. Overall, the interviews provided the rich detail of everyday life, as well as the former residents' perceptions of that life. Interview material also allowed researchers to gain some understanding of people's motivations and attitudes, material that cannot be gained from the census. On the other hand, census data provided aggregate information that was invaluable in understanding the racial and ethnic composition of Buxton and the range of occupations, class structure, family size, income, and unemployment there. Taken together, we believe that the major sources have provided an abundance of materials from which to reconstruct the Buxton experience.

1

MUCHAKINOCK: BUXTON'S HISTORICAL ANTECEDENT

IN the late nineteenth and early twentieth centuries, thousands of people arrived in Iowa, seeking work in one of the state's nearly four hundred coal mines. Many newcomers, foreign-born and native-born alike, would find employment in small coal camps like Smoky Hollow, Foster, and Enterprise, while perhaps an equal number would settle in larger, incorporated communities such as Seymour, Madrid, and Des Moines. Some newcomers would head for Muchakinock, an unincorporated coal town in southern Mahaska County. Founded in the late 1860s, Muchakinock was originally little more than a trading center, but in 1881 the camp and the surrounding coal mines came under new management. In that year, the Chicago and North Western Railroad purchased the property, known as the Consolidation Coal Company, and, because of labor difficulties, began to import southern blacks as strikebreakers. Muchakinock, or Muchy as the camp was commonly known, would soon become the state's largest unincorporated coal mining community and a major settlement area for blacks.

While Muchakinock was significant in its own right, it was also closely tied to the development of an even larger community. In 1900, most residents of Muchy would move to Buxton, a new camp located to the southwest in Monroe County. Because of the carryover of population and the continued presence of the Consolidation Coal Company, Muchakinock provides the historical antecedent for Buxton. The roots of the latter community go back in almost every detail to the Muchy camp. There blacks and whites lived and worked side by side, most of whom would later work in Buxton. Blacks established churches, lodges, and businesses that they later relocated in Buxton, and other ethnic groups, particularly Swedes, also settled in Muchy and later moved on to the new camp. In Muchakinock, moreover, the Consolidation Coal Company firmly established itself as the largest and most influential coal company in the state. So, because of the close social and economic ties between the two communities, the story of Buxton properly begins in Muchakinock.

While the takeover by Consolidation in 1881 marked the most significant point in Muchakinock's history, the community had begun to attract coal miners almost a decade earlier.[1] In 1873 two brothers, H. W. and W. A. McNeill, moved from Coalfield, where they had mining interests, to the Muchakinock valley. They immediately organized the Iowa Central Coal Company and in the process absorbed three local firms, the Hardin, the Mahaska, and the Southern coal companies. They also bought out the Black Diamond Mine in Monroe County in 1875 and then reorganized all their holdings into the Consolidation Coal Company. Company officials listed their capital at $500,000. Two years later they purchased the Mahaska County Coal Company, making their firm the largest coal mining operation in Mahaska County. In 1878, Muchakinock contained several hundred inhabitants. Community businesses included a general store and a drugstore. The Welch Congregational Church, organized before 1870, was the only church. A local observer noted: "The miners are generally quiet and industrious, and there is not much drinking, although the town has a saloon." Muchakinock's first railroad began operation in 1873, when the Iowa Central Railroad ran a spur line from Givin, a nearby coal camp, to Muchakinock. Consolidation sold most of its coal to railroads in northern Iowa and Minnesota.[2]

The expansion of Muchakinock in the 1870s was the natural outgrowth of larger, more diversified economic changes within the state of Iowa. In the decade following the Civil War, Iowa experienced extraordinary economic growth. Officially opened for white settlement in 1833, Iowa's population had soared to 1,624,615 by 1880.[3] The state's rich, black soil attracted thousands of settlers from the Northeast, particularly Pennsylvania, Ohio, and Indiana, and immigrants from northern and western Europe also settled in Iowa in ever-increasing numbers. In the 1850s, Iowans became caught up with railroad fever as eastern Iowa promoters worked to organize railroad companies. The Civil War temporarily disrupted railroad building, but by 1870 four railroad lines crossed the state. A few years later, as settlement pushed into northwest Iowa, most of the state lay blanketed with farms and small towns.[4]

As a result of its rapid economic expansion, the Hawkeye State had reached an important milestone by the 1870s. Within only three decades, Iowa had passed from the first primitive stage of pioneering to economic stability and well-being. Iowans had proved themselves to be excellent agriculturists, and by the 1870s most farmers had long since reached the commercial stage of production. Railroads provided Iowa farmers with access to eastern markets and to the international markets beyond. At the same time, the railroads, in connection with expanding urban populations in eastern Iowa, made possible the beginning of numerous industrial operations.

The coal mining industry developed as a result of Iowa's railroad expansion. Spurred on by the railroads' need for coal, local entrepreneurs began in the 1870s to explore substantial coal reserves in central and southern Iowa. Reports of Iowa coal reserves had been published as early as 1835, following a survey by a United States Geological Survey team. In that year Lieutenant Albert Lea led an expedition to investigate the newly acquired Black Hawk Purchase. Lea reported that large coal deposits lay along the Des Moines River near the Raccoon Fork. Iowans had already been conducting limited mining operations in Van Buren, Scott, Muscatine, and Jefferson counties in eastern Iowa, but the first major development of Iowa coal took place in Des Moines in the 1860s under the direction of Wesley Redhead. Other companies soon followed, and in the 1870s operators developed mines in most parts of the state's coalfields.[5]

Before 1870, Iowa coal was utilized primarily for heating homes and businesses, but the expansion of Iowa's railroad mileage between 1870 and 1920 provided the major impetus for the development of its coal deposits. Railroads provided a constant demand for coal, while previously demand had been only seasonal. As a result, in 1885 the state mine inspector reported that Iowa contained 489 underground coal mines.[6] Railroads followed the practice of creating "captive mines," whereby they took control of a particular coal mine and consumed all the coal produced in it. This arrangement guaranteed the railroads a steady, year-round supply of fuel. Thus, from 1870 the well-being of Iowa's coal mining industry was tied closely to the policies and practices of its railroads.

As more coal mines opened, coal camps also began to appear. Because the mines were usually located away from population centers, operators had to provide housing for mine workers and their families. Central and southern Iowa were soon covered with dozens of small camps, such as Carbonado, Keb, Pekay, and Colon, most of which appeared bleak and dismal to residents and outsiders alike.[7] Coal operators usually developed the camps quickly, giving little attention to their physical surroundings or to the quality of house construction in them. The typical coal-camp house was a one-story, four-room structure. Because operators did not expect the coal mines to exist for more than eight to ten years, they did not worry about the maintenance of company houses, nor did they bother planting grass, trees, or shrubs. Beyond these houses, the typical coal camp included only a company store, a union hall, a tavern, and a grade school, so camp residents enjoyed few social and leisure-time activities.[8]

Muchakinock was not destined to be a typical coal camp, however. In 1881 the McNeill brothers sold the Consolidation Coal Company to the Chicago and North Western Railroad for $500,000. Marvin Hughitt became the first president and John Emory Buxton the first general superintendent under the railroad's ownership. At the time of purchase, Consolidation was operating three mines, and the railroad soon constructed a line from Belle Plaine to the Muchakinock coalfields and a depot and rail yard in the camp itself. During its twenty years in Muchakinock, Consolidation was the major supplier of locomotive coal for the Chicago and North Western Railroad.[9]

Soon after its takeover by the Chicago and North Western, Muchakinock became known as a predominantly black community. This designation grew out of action taken by Consolidation officials in the area of labor recruitment. In 1881, Consolidation changed the method of paying its employees, and the miners reacted to the change by striking. During the strike, Consolidation officials sent Hobe Armstrong, a local black businessman and farmer, to Virginia to recruit black mine laborers. It is not known whether Armstrong informed the new recruits that a strike was in progress, but in effect the blacks were hired as strikebreakers. Armstrong succeeded in recruiting a large number of blacks, particularly from the Charlottesville and Staunton areas, and the company then shipped them to Iowa in railroad cars. The blacks had no experience as coal miners. It is unclear whether Consolidation continued to import southern blacks as strikebreakers during later strikes, but apparently the company did decide that southern blacks constituted a good, cheap form of labor, as from 1881 on it recruited them as regular mine workers, placing them with experienced white miners until they could function independently. From 1881 on, Muchakinock contained a sizeable black population, al-

Store clerks, company miners, and local residents gather on the steps of the Consolidation Coal Company's store in Muchakinock in the 1890s. (*Courtesy of Iowa Mines and Minerals Department, Des Moines*)

though the proportion of black residents changed considerably between 1881 and 1900. In 1885, blacks accounted for approximately 66 percent of Muchy's total population, but ten years later, the figure had dropped to around 37 percent.[10] In the view of neighboring residents, however, Muchy remained a predominantly black community throughout its existence.[11]

Although Muchakinock contained the state's largest concentration of blacks, it was not the first Iowa coal camp to import blacks as strikebreakers. In February, 1880, Henry Miller, owner of the Albia Coal Mine in Monroe County, had imported blacks from Missouri to break a strike at his mine. The blacks faced a hostile reception the night they arrived, being shot at by local residents, but a company of militia from Albia restored order before any casualties resulted. During the same year, the Whitebreast Coal and Mining Company in

Lucas County also imported southern blacks as strikebreakers. Faced with a strike and unable to reach an agreement with their workers, Whitebreast officials sent agents to West Virginia to recruit black laborers for the mines near Cleveland. To house the black workers, the company built a new camp, called East Cleveland, adjacent to the original camp. Within a short time, however, the mines shut down and most of the blacks moved to other northern states or returned to West Virginia.[12]

The presence of a large number of blacks at Muchakinock marked a sharp departure from Iowa's nearly all-white population. Iowa had been settled by immigrants from the British Isles and from northern and western Europe, as well as by white settlers from the Northeast and Middle West whose ancestors had earlier emigrated from Europe and the British Isles. The number of blacks in Iowa had remained low. Seven years after Iowa officially opened for settlement, the 1840 Iowa territorial census included 153 free blacks and 18 slaves out of a total population of 43,112.[13] The few slaves who appeared in Iowa before the Civil War were brought here by their southern masters and resided in the extreme southeastern part of the state. Officially, slavery never existed in Iowa, for two reasons. First, the Northwest Ordinance outlawed slavery in the Northwest Territory, and this applied to Iowa Territory (and later to the state of Iowa) as part of the Louisiana Purchase. Second, with the adoption of the Missouri Compromise in 1820, Congress determined that the area north of latitude 36° 30′ (except Missouri) was to be free of slavery. Iowa's territorial legislature also enacted laws that hindered the movement of free blacks into the area. In 1839 and 1840 it passed two laws that later became known as Iowa's Black Code. The first law stated that "no black or mulatto" would be permitted to settle in Iowa without possessing a certificate of freedom and the ability to post a bond of $500 indicating that the black would not become a public charge. The second law, passed in 1840, prohibited marriage between blacks and whites. These actions reflected Iowans' strong belief that blacks should be kept out of Iowa Territory.[14]

Four years later Iowans again faced the question of legal rights for blacks when they set out to write a state constitution. Although the vast majority of Iowans, like most white Americans, believed in white racial superiority, they also opposed slavery. In fact, many Iowans, particularly Quakers and Congregationalists, took an active part in the underground railroad. The granting of political and social equality to blacks, however, was viewed as a separate matter. When Iowans assembled in 1844 to draft their state constitution, Henry County delegate George Hobson introduced a petition asking his fellow delegates "to secure to people of color all such rights and privileges, civil, social, moral and educational, under the same circumstances and upon the same conditions as are secured to others." The convention quickly referred the petition to a committee. After several days, the committee reported that, while they accepted the premise of equality for all mankind, this was "a mere abstract proposition." They believed that because government was an artificial association it might, through its constitution, modify an individual's natural rights. They could "never consent to open the doors of our beautiful State" to the black man. In its final form, the Constitution of 1844 (which was essentially embodied in the Constitution of 1846) excluded black men from voting and from serving in the state legislature and the state militia. During the same period, black children could not attend public schools.[15]

After the Civil War, however, Iowans reacted differently to the presence of blacks in the state, making several changes that resulted in legal equality for black men. In 1868, Alexander Clark, a black businessman from Muscatine, challenged Iowa's segregated schools. Clark initiated litigation that resulted in several state supreme court decisions whereby his daughter and other black children were permitted to attend the state's public schools. The same year, Iowans passed a referendum that removed the word "white" from the state constitution, granting black men the vote. At the time, only five New England states extended suffrage to blacks. Later, in 1880, Iowans removed the word "white" from the qualification for serving in the state legislature. With this move, black men in Iowa achieved equal political rights with white men. Finally, five years later the state legislature passed the Civil Rights Act of 1884, which stated that "all persons within this state shall be entitled to the full and equal enjoyment of the accommodations, advantages, facilities and privileges of inns, public conveyances, barber shops, theaters and other places of amusement." In 1892 the legislature enlarged the list to include eating places and bathhouses.[16]

After the Civil War, the number of blacks in Iowa rose substantially. Many newly freed blacks traveled up the Mississippi River, settling in the river communities of Burlington, Davenport, Clinton, and Dubuque. Some also traveled up the Missouri River, settling in Council Bluffs and Sioux City. From these cities, blacks hired on as steamboat deckhands. At Keokuk, because of rapids in the river, steamboat cargoes had to be unloaded, carted overland for a short distance, and then reloaded. Before the Civil War, Irish immigrants handled this work, but after the war blacks were hired. In eastern Iowa, blacks also worked in railroad construction. In some cases, these same men later worked for the railroads as porters, waiters, and cooks. Frequently blacks also moved from their initial employment as deckhands into service areas. Some blacks became barbers and waiters, while others opened restaurants and laundries; some worked on construction projects. Industrial opportunities in the Davenport area also attracted black labor in the 1880s, but few blacks purchased land and became farmers.[17] Of all the occupations open to blacks in Iowa after the Civil War, however, coal mining attracted the greatest number. During the decade of the 1870s, when Iowa's coal mining industry was expanding significantly, blacks responded to that opportunity, particularly those living in Missouri. By 1880, 9,516 blacks resided in Iowa, an increase of 38 percent over 1870. By 1890, the number had risen to 10,685, and by 1900 it reached 12,693.[18]

With the takeover by the Chicago and North Western Railroad in 1881, Muchakinock entered into a period of rapid growth. In 1885, Polk's *Iowa State Gazetter and Business Directory* contained the following description: "[Muchakinock] contains Baptist, Methodist, and 2 Swedish churches, public schools, and have [sic] extensive coal mines, coal being the only shipment. 4 stages daily to Oskaloosa, fare 25 cents. Population, 1,200. Mail daily. J. E. Buxton, postmaster." The community also included a meat market, a drugstore, a general store, a saloon, a blacksmith's shop, and an opera house, as well as an office for two company physicians. Consolidation also maintained both a pay office and a general office in the camp.[19]

Between 1885 and 1895, Muchy's population more than tripled, reaching 3,844 people by 1895. During the 1890s the community attracted numerous new businesses, including several restaurants, two saloons, and a telephone office. Minnie B. London, who moved to Muchakinock in 1891, remembered that the camp then included a town hall, four grade schools, a mine superintendent's home (occupied by the John E. Buxton family), and four churches. Two churches, the African Methodist Episcopal and the Union Valley Baptist, were attended by blacks, while the Swedish Lutheran and the Swedish

Muchakinock's main street as it looked in the 1890s. (*Courtesy of Iowa Mines and Minerals Department, Des Moines*)

Methodist churches were attended by whites.[20]

The management of the Consolidation Coal Company was handled by the company's general superintendent, John Emory Buxton. Born in Middletown, Vermont, in 1840, Buxton owned a dairy farm there. In the late 1870s he moved to the Middle West and became a tie buyer for the Chicago and North Western Railroad. A few years later, Consolidation appointed him the general superintendent of the Muchakinock mines. By 1887, Consolidation had four mines in operation, Nos. 1, 2, 3, and 5. The company employed approximately 500 men, 350 of whom were black. Ten years later, the Iowa state mine inspector reported that Consolidation operated four mines, Nos. 6, 7, 8, and 9. The inspector noted that "No. 9 is a new plant, the shaft being sunk and equipped with all modern improvements within the past year." Since all mines were located some distance from the camp, trains transported the miners to and from the mines. In 1897, Consolidation employed approximately 600 men.[21]

Although Muchy was a company town, residents had a choice regarding housing. The company owned some houses that it rented to miners, but they could also build homes on land rented from the company. Herman Brooks remembered that many families, both black and white, owned their homes. These were located on half-acre lots that families rented from the company for one dollar per month. Brooks recalled that many families raised chickens, cows, and hogs. His family also raised ducks, geese, and pigeons. Brooks related that "everyone raised and butchered their own meat." Most lots contained caves or root cellars where people stored their food. Bessie Lewis's family home was located on a large lot, where they kept chickens, pigs, and a cow. Lewis's mother raised a large garden from which she canned fruit and vegetables. She also canned large quantities of meat. The family had a well in their yard and procured their water by lowering a bucket into it. Apparently not every family had a well, however, as Jeanette Adams recalled drawing well water and carrying it a long distance to her home. Some black families—including the George Willises, W. E. Andersons, James Colemans, John Joneses, O. C. Carters, and the Steels—had homes located on two-acre lots.[22]

Although interviews and other data indicate that blacks and whites lived side by side in Muchy, the camp did contain a few subdivisions. Bessie Lewis remembered that there was some ethnic division, as the camp had a section known as Swedetown, which contained Swedish miners and their families. The town also contained two smaller areas or subdivisions known as Unionville and Newton.[23]

The great majority of black people who lived in Muchakinock had migrated from Virginia. In 1885 the state census showed that 85 percent of the blacks there had been born in Virginia. Ten years later that number had dropped to 51 percent.[24] Throughout the 1880s and 1890s, Consolidation continued to recruit black workers from Virginia, and even as late as 1898 Virginia blacks were still arriving in Muchy.[25]

The Albert Rhodes family was typical of black families who migrated from Virginia in the mid-1880s. The family had formerly lived in Charlottesville, Virginia, where Rhodes and his wife, Margaret Elizabeth, had eight children. Two more children were born after the family arrived in Iowa. Rhodes's oldest brother, Addison, along with his wife and two children, had settled in Muchy five years earlier, and he had returned to Virginia and persuaded his brother, Albert, to join him at Muchy, telling him that it was possible to make more money in Iowa. Eventually, through Addison's efforts his parents, grandparents, and three brothers joined him. Addison and his two brothers went to work for Consolidation as miners. For Albert Rhodes the move was a wise one, as he later had the opportunity to operate both a livery stable in Muchakinock and a hack service from Muchakinock to Oskaloosa. His daughter, Bessie Lewis, remembered that he made four trips a day, two in the morning and two in the afternoon. She added that her father "had horses and all kinds of buggies and rigs . . . so he carried a lot of people over there in his vehicles."[26]

In their trips to Virginia, Consolidation agents also recruited single black men and at least a few single black women. Mattie Murray remembered that both her parents had been recruited to work in Muchakinock. They married after settling there. Murray's father, Tandy Bradley, arrived in Muchakinock in 1889. Bradley, born in Charlottesville, Virginia, believed that he was coming to work in an Iowa gold mine. Murray remembered her father telling her about his move to Iowa. She related:

> The man went down there that had the Muchakinock coal mines and gathered up all them young boys and brought 'em here to dig coal. My father couldn't read or write. He didn't know his name when he seen it. He had never gone to school. All he knew was to work on a farm down there. And Consolidation's agents came down there and gathered up all of them, my father and mother, and they brought all of them people out here. They brought the young women out here to work for different people.[27]

Murray's mother, Ella Fry, was eighteen years old when she arrived in Muchy. The Bradleys had five children in Muchakinock and twelve more after they moved to Buxton. Bradley worked as a coal miner for Consolidation for many years. Murray remembered that other relatives were also recruited by Consolidation.[28]

John Chapman was another single black teenager recruited by Consolidation in the early 1880s. Chapman lived in Staunton, Virginia, where he worked in a bakery. After arriving in Muchakinock, Chapman married a young woman who had migrated to Muchy with her parents from Charlottesville, Virginia. The Chapmans had seven children in Muchy, and later the entire family moved to Buxton.[29]

A substantial number of blacks migrated to Muchakinock from Missouri. It is not known if they did so independently or if Consolidation recruited there. The Joseph Brooks family, for example, arrived in Muchakinock during the 1880s. Born in Mississippi, Brooks moved to Missouri and worked in Kansas City for a time before moving to Lexington. There he met his wife, Jennie Robinson. The Brooks had six children, three sons and three daughters. Brooks worked as a miner in Muchakinock and later in Buxton.[30]

Some residents of Muchy had also immigrated from European countries. Sometime during the 1880s, Charles Erickson traveled to Iowa from northern Sweden. He had worked as a miner in Sweden and wished to continue that occupation in the United States. Friends in Muchy influenced his decision to locate there. When he arrived, he met and married a young Swedish girl, Josephine Carlson, who had also immigrated there with her family. Josephine Carlson had come from Gotëberg, Sweden where her father had been a fisherman. The Carlson family had first immigrated to Pennsylvania because a son-in-law had settled there and had become a contractor for railroad construction. He then encouraged his wife's family, including her parents and siblings, to join him in the United States. Alex Erickson related the following about his uncle:

> [He] had become a contractor. . . . He would take contracts that he'd put in the railroad for so many miles. . . . Then he hired Swedes that came here, couldn't talk or anything, and he bossed them, and he got a lot of work out of them because they were big and strong, and he went ahead and made money. After he had made so much money, he got my grandpa to come [to Pennsylvania] and then my mother came here and her sister, and then my grandmother came here. They were all in Pennsylvania, and then they came to Iowa.[31]

Charles Erickson worked as a miner for Consolidation in both Muchakinock and Buxton. He and his wife had three children, two daughters and a son. Their son, Alex, worked as a miner for Consolidation for over thirty years and their daughter, Agnes, worked in the company's business office.[32]

The George and Anna Mital family also emigrated from Europe to Muchy. Although of Slovakian ancestry, George Mital was born in Germany and Anna in Yugoslavia (probably Croatia). Both individuals came to the United States as young people. Anna was sixteen at the time she immigrated with her family to Lehigh, Pennsylvania, where her father worked in the steel mills. The couple met and married there and then moved to Minneapolis, where George worked in a sawmill. Sometime in the mid-1880s, George Mital's brother-in-law, John Halansky, traveled from Pennsylvania to Muchakinock, where he went to work in the coal mines. Apparently satisfied with his new job, he sent George a train ticket, urging him to come to Muchakinock. By 1888, when George took out his final naturalization papers, he and his wife were living in Muchakinock. In 1892 their son Andrew was born there.[33]

As these family experiences indicate, both southern blacks and European immigrants found family and ethnic ties to be extremely important, perhaps even crucial, in determining that they would locate in Iowa. Certainly the decision to move to Muchy was made easier with the knowledge that some family members or friends had already made the move and stood ready to offer assistance. Not only did relatives and friends inform the newcomers of job opportunities, but also they frequently provided money for their transportation as well as an initial place to live. Throughout the 1880s and 1890s, Muchakinock's residents continued to urge relatives and friends to join them in Iowa. Many wrote letters back home telling about the steady work and the good pay. The *Bystander* frequently reported on trips taken by Muchy residents to visit family members in Virginia. No doubt blacks used these trips to urge family members and friends to come west and work for Consolidation. Foreign-born residents also wrote home encouraging others to come to Muchakinock and work in the mines. Overall, family and ethnic ties played a vital role in the development of the camp.

Throughout the 1880s, Muchakinock's black community became increasingly visible and cohesive. Black coal miners constituted the largest occupational group, but blacks also worked in other occupations. Moreover, black families built churches and lodges, and black children attended the integrated public schools. Although black business and professional people were the most influential citizens, Muchy's black coal miners also formed organizations and played important roles in shaping the black community. Taken together, the black groups constituted a strong and cohesive community that included political, social, economic, and religious institutions.

Muchakinock's black population included a substantial number of business and professional people. The most prominent black businessman was Hobert A. Armstrong, known widely as Hobe. Armstrong played a major role in Muchy's economic life and became one of the area's wealthiest men. It is not known why Armstrong settled in Muchakinock. Born in Knoxville, Tennessee, in 1850, Armstrong was orphaned when he was twelve. After his parents' death, he went to live with a white man, a Dr. Perdue. Eventually Perdue moved to Kansas, taking Armstrong along. Perdue purchased a farm there and it later became the property of Armstrong. Armstrong moved to Muchakinock in the mid-1870s and soon married the daughter of a local German farmer. Hobe and his wife, Anna, had eleven children.[34]

Soon after arriving in Muchy, Armstrong developed close business ties with the Iowa Central Coal Company, arranging to purchase mules for the coal company and contracting to haul coal as well. He bought the mules on consignment and then sold them to the company. A short time later, Armstrong opened a meat market in Muchakinock. Although Armstrong's meat business was not company owned, miners paid for their meat purchases with company scrip, which was then processed through the company's business office. Armstrong's business ties carried over to Consolidation when it purchased the mines, and he also began buying farms in the Muchakinock area. His son-in-law, Carl Kietzman, estimated that Armstrong owned seventeen or eighteen farms in southern Mahaska and northern Monroe counties. Armstrong also raised Thoroughbred horses.[35]

The view that emerges of Hobe Armstrong is that of a private but generous person whose time was spent caring for his business interests. Kietzman, who greatly admired his father-in-law, described his generosity, recalling that Armstrong did not belong to any church but made contributions to all the churches in the area. Kietzman believed that Armstrong was the type of man "who would help anyone." Moreover, he described Armstrong as a temperate man who rejected alcohol his entire life.[36] Another Muchy resident, Minnie B. London, also recorded her views of the man, particularly his belief in interracial marriage. London wrote in her memoirs:

> H. A. Armstrong who owned and operated the meat markets in both camps acquired considerable wealth, and I might say was a man of strong convictions and advocated that the solution of the race problem could be accomplished by inter-racial marriage and in his own life practiced what he preached.[37]

According to many Buxton residents, Armstrong also strongly encouraged his children to marry whites, which they did.[38]

Other residents remembered that Armstrong accumulated considerable wealth. Clayborne Carter recounted that Armstrong had the reputation of being one of the largest landowners in Mahaska County and was worth half a million dollars. Earl Smith, a lifelong resident of the Buxton area, recalled that Armstrong also had rental houses in Buxton: "I don't know where the houses come from, but over there close to where he lived he had several houses that he rented, and then he started on buying these farms. . . . About every year he'd buy another farm, kept doing that and . . . when he died, he had a lot of them." Smith remembered that Armstrong frequently purchased land around the coal mines.[39]

To many Muchakinock and Buxton residents, Armstrong was best remembered as the man who recruited the first blacks to work for

Consolidation. At the time Armstrong went south to hire blacks, he had considerable contact with the local coal company. When asked why the company hired Armstrong for this purpose, his son-in-law replied: "Well, he had a way; he was a good conversationalist. He was good natured." Kietzman added that Consolidation officials had "liked [Armstrong] and he favored them." He remembered that Armstrong "just hired them people down there, and they were poor people, and [the company] would pay the moving expenses, . . . and they'd pay them so much per week, and they'd learn to mine coal."[40]

As well as attracting black coal miners and businessmen, Muchakinock also attracted a small nucleus of black professionals. George Woodson, a black attorney, settled in Muchakinock in 1896. Born in Wytheville, Virginia, where both parents were ex-slaves, Woodson graduated from Petersburg Normal University in Petersburg, Virginia, in 1890. He then enlisted in the army and served in the West. A few years later he entered Howard University, where he received a law degree. He graduated at the head of his class in 1895, receiving the class medal for oratory. The following year Woodson moved to Muchakinock, where he practiced law. He soon became active in professional organizations, serving as vice-president of the Mahaska County Bar Association. Shortly after his arrival in Muchakinock, the editor of the *Bystander* wrote: "George Woodson [is] the rising young attorney-at-law who within less than one year has several times met the giants of legal minds at the Oskaloosa bar and surprised the bar by his proficiency as a barrister." Woodson soon became active in the Republican party, frequently serving as a delegate to both county and state conventions. In January 1898, Woodson ran for the position of file clerk in the Iowa Senate but was defeated by only one vote. Later he ran unsuccessful for county attorney of Mahaska County. Woodson was an extremely popular public speaker and frequently appeared at Fourth of July celebrations, church affairs, and school events.[41]

Other black professionals in Muchakinock included teachers, ministers, and a druggist. In the early 1890s, L. A. Wiles came to Muchakinock from Virginia, one of several professionally trained educators attracted to Muchy in that decade. Born in Florida, Wiles attended Hampton Normal School for four years. He taught school in Virginia for several years and then came to Muchakinock to teach. He remained there for several years and then took a position with the Census Bureau in Washington. Frances Brown and S. Joe Brown, sister and brother, came to Muchakinock in the mid-1890s. Miss Brown had graduated from Ottumwa High School and had attended

George Woodson, who practiced law in both Muchakinock and Buxton. (*Courtesy of Iowa State Historical Department*)

Lincoln Institute, the state normal school for black teachers in Jefferson City, Missouri. She taught several years in Missouri before coming to Muchakinock. She first served as the principal of the Muchy public schools but soon resigned because of poor health and became an elementary school teacher in the system. Her brother succeeded her as principal. S. Joe Brown had also graduated from Ottumwa High School and then had taken "a full scientific course" at the University of Iowa. In 1898, Brown left Muchakinock to take a position at Bishop College in Marshall, Texas. Another black professional, the Rev. C. R. Brookins, served for several years as both minister of the African Methodist Episcopal Church and school principal. Brookins had been born in Virginia. B. F. Cooper, a pharmacist from Clinton, came to Muchakinock in the 1890s and opened a drugstore. Cooper played a leading role in the community's business affairs, as well as managing the local baseball team, the Unions.[42]

From early in Muchy's history, the camp's black coal miners also played important leadership roles. The most visible evidence of that leadership lay in the existence of "the Colony," a society formed by blacks for mutual economic protection. Organized in 1881, the society's main purpose was to provide benefits for medical care and burial expenses. Both Clayborne Carter and Minnie B. London remembered that Consolidation initiated the founding of the Colony by suggesting the idea to the black miners. Carter and London recalled that the Mahaska County supervisors were fearful that the southern blacks would have a difficult time adjusting to Iowa's cold weather and might therefore experience considerable sickness and death. To guard against additional expenses for the county, the supervisors suggested to the company that the miners form a mutual protection society. The company agreed and required the miners to participate. Married men paid dues of one dollar per month and single men paid fifty cents a month. The Colony paid 80 percent of the dues to a physician for medical care and placed the remainder in a burial fund. Officers reported in 1886 that the Colony had a balance of five hundred dollars. When a miner became ill, he received three dollars a week for compensation; in case of death, the deceased's family received forty dollars. Even though the Colony had been suggested by outsiders, Muchy's black miners found the organization useful and supported it enthusiastically. Once founded, the group elected officers and provided their own leadership. They continued the organization until Muchy residents relocated in Buxton. C. R. Foster, a weighmaster for Consolidation, served for many years as its president.[43]

Although the Colony's main function was mutual assistance, the association apparently played a variety of roles. Informally, the group provided a means of self-government for the camp's black population. In October 1883, the *Oskaloosa Weekly Herald* reported that the officers of the Colony "met several times during the week and fined some of the boys for getting too much free water on board and consequently acting disorderly." Throughout the 1880s and 1890s, the Colony continued to provide medical care and burial assistance for its members and to promote good behavior and frugal living.[44]

Churches and lodges constituted Muchakinock's main black community institutions. According to the *Bystander* and other area newspapers, the camp's two black churches, the African Methodist Episcopal and the Union Valley Baptist, served as centers of both religious and social life. Together the churches sponsored numerous religious, educational, and social clubs. One of the largest groups sponsored by the Baptist church was the Baptist Young People's Union. The BYPU met monthly on Sunday afternoons to hear devotions and literary works. In March 1897 the *Oskaloosa Daily News* reported that the BYPU had met for devotional readings and presentations from the works of Longfellow.[45]

Muchakinock was also the home of a variety of black lodges, including the Odd Fellows, the Household of Ruth, the Masons, the Virginia Queen's Court, and the Knights and Daughters of the Tabernacle. In July 1896, the Odd Fellows held their annual session of District Grand Lodge No. 30. The two-day affair, which attracted delegates from various parts of the state, included entertainment, a parade, and the installation of new officers. Muchakinock members rented three halls in addition to their own lodge to accommodate visitors. The highlight of the convention was the parade, which ended at the fairgrounds. George Woodson gave the main address.[46]

In the 1890s, Muchy was served by two black newspapers, the *Negro Solicitor* and the short-lived *Muchakinock State*. George Taylor, a black man, edited the *Solicitor* in nearby Oskaloosa. The *Muchakinock State* was founded in December 1897 by seven men in Muchakinock and apparently lasted for several months. J. Ed White served as editor-in-chief, and the following men served as associate editors: C. R. Foster, a weighmaster for Consolidation; Nathan B. McDowell, a saloon keeper; Jacob Wilson, a coal miner; Martin Crowder, a bookkeeper; and George Childs, a laborer. It is not known whether the editor was black, but Foster, McDowell, and Wilson were. The work of these men in founding the newspaper provides testimony that blacks from all occupational levels played strong leadership roles within the black community. Moreover, law enforcement in Muchakinock was carried out by a black constable and a black justice of the peace.[47]

The Young Men's Christian Association, an organization that appeared late in Muchy's history, also presents evidence of a strong black community. In 1899 the *Bystander* announced that a YMCA would be organized in Muchakinock. Although the organization was short-lived (because of the population's relocation in Buxton), it is apparent from black residents' statements that they viewed the "Y" as an institution that would benefit their race in many ways, as well as provide for greater racial solidarity. On June 30, 1899, the Muchy reporter for the weekly *Oskaloosa Herald* wrote: "Muchakinock is to have a Y.M.C.A. which will perhaps be the only one in the state composed entirely of Negroes. We lead, others follow." A month later, the same scribe wrote: "about fifty men [were] present of all ages ranging from fourteen to sixty and all expressed a desire to have the meeting continued and a permanent association organized." By September 1899 the Muchy "Y" was organized and holding monthly meetings.[48]

The participation of Muchy blacks in all phases of community life touched a prevailing theme running through the community's entire history: a desire by blacks to educate themselves, to advance themselves economically and socially, and to uplift themselves as a race. Black leaders like George Woodson, the Reverend Mr. C. R. Brookin, Rev. William Williams (pastor of the African Methodist Episcopal Church), and Frances and S. Joe Brown repeatedly admonished fellow blacks to improve themselves in every way. Black leaders continually urged black children to attend Sunday school and church more frequently, to attend public school more regularly, and to study more diligently. In talks to black adults, black leaders urged them to be better Christians and to be more regular churchgoers. Moreover, they urged adults to live frugally and to save their money. In May 1897 the Muchakinock reporter for the *Bystander* included the notice: "Young men who labor every day, improve your opportunity by attending Rev. T. L. Griffith's night school." [49]

Local blacks also sought to uplift members of their race by sponsoring the appearance of black individuals who had achieved success in the outside world. In February 1899, Booker T. Washington presented a lecture in Muchakinock, an event heralded by the entire community. The local newspapers urged everyone to attend. As a result, the entire town closed down so that all residents might hear Washington's lecture. According to the *Bystander*'s report, Washington "advised the accumulation of money and property, the giving of trade to boys and girls as well as good education." The *Bystander* noted that Muchakinock would never forget the speech "which already opened many hitherto blinded eyes." Apparently Washington's speech had some effect, as a few days later George Woodson called a meeting to organize a Muchakinock Banking Club, presumably a group that would encourage residents to begin savings accounts. The group elected Woodson president and S. Joe Brown secretary.[50]

Muchakinock blacks also involved themselves in promoting college education for black young people. In July 1899, while fulfilling a speaking engagement in Mt. Pleasant, George Woodson assisted in organizing the Iowa Negroes' College Fraternity, an honorary society to which only black college graduates could belong. The group started with six charter members and expressed the hope that they would soon include black college graduates from all over the state. The *Bystander* noted: "This is perhaps the first organization of its kind in this section of the country and certainly speaks well for Negroes of this state." On the local level, Muchakinock residents worked together to help a local black youth pay his college expenses. In May 1899, Edward Carter graduated from Oskaloosa High School, the only black member of his class. Muchy residents knew that Carter hoped to attend the University of Iowa, so they went to work to help finance his college education. In July a benefit was held at the Muchy Opera House, where Carter was honored and presented with donations from local residents. Townspeople expressed great interest in Carter's work at the University, and the *Bystander* reported faithfully on Carter's home visits.[51]

While Muchy's black population took part in many organizations, they reserved their greatest enthusiasm for the Republican party. The *Solicitor* and the *Bystander* were filled with articles covering the activities of Iowa's Republican party, as well as frequent references to the national Republican party. Muchy blacks not only participated in the party but at times attempted to bring about changes that would give a greater voice to black members. Given Mahaska County's large black population, Muchakinock's black leaders rightfully believed that they should have some influence in the county's Republican organization and some representation in state government. At one time, Mahaska County's black delegation to the state convention offered a resolution urging that body to enlarge its central committee to include a black member. The convention tabled the resolution.[52]

Blacks from Muchakinock did not always have the political backing of blacks in other Iowa counties, particularly Polk County, the state's most populous county and the location of the state capital. As mentioned earlier, in January 1898, George Woodson ran for file clerk in the Iowa Senate and was defeated by only one vote. A short time later the *Solicitor* carried an article by Charles D. Ruff, a black Des Moines resident, entitled "Negroes Turned Down." In general, Ruff criticized Iowa Republicans for relegating Iowa blacks to menial positions in the party. Specifically, Ruff chastized the black Polk County Republicans for refusing to support Woodson's bid for the file clerk's position. Ruff noted that both the *Solicitor* and the *Des Moines Register* had given Woodson "complimentary notices," while the *Bystander* had failed to do so. Ruff also criticized the Polk County black Republicans for frequently supporting poorly qualified black candidates while bypassing talented men like Woodson.[53]

But Woodson's defeat did not dampen the enthusiasm of Muchy's black residents for political activities. Woodson continued to play a highly visible political role, running for county office and serving as Mahaska County delegate to many state Republican conventions. In August 1899 he was elected chairman of the Mahaska County Republican Convention. Another Muchy resident, Hobe Armstrong, was mentioned from time to time as a possible candidate for the Iowa General Assembly, but it is not clear whether he was ever nominated.[54]

Although Muchakinock blacks were often unsuccessful in their quest for county and state political offices, they did dominate at the township level. In July 1899 the *Bystander* reported that East Des Moines Township (which included much of Muchakinock) had held its caucus and that blacks had been elected to all offices and had comprised all the delegates to the Republican county convention. The caucus expressed the hope that two members, C. R. Foster and George Woodson, would be elected as delegates to the Republican state convention. The *Bystander* concluded: "The race will be better represented in Mahaska than any other county convention in the state."[55]

Although Muchakinock early developed a reputation as a black community, it also contained a wide variety of other ethnic groups. In June 1883 the Muchy reporter for the *Oskaloosa Weekly Herald* reported that "there are about 1,200 inhabitants in the place: 700 or 800 of these are colored, and the rest are mostly Swedes."[56] In 1885 the state census showed that Muchy contained 795 blacks and a wide

Table 1.1. Birthplace of Foreign-Born Coal Miners, Muchakinock, 1885

Birthplace	Number	Percent of Total Population
Sweden	72	6.0
England	71	5.9
Wales	51	4.3
Ireland	25	2.0
Scotland	23	1.9
Germany	8	0.6
Hungary	7	0.6
Norway	3	0.2
France	1	0.1
Total	261	21.6

Source: Iowa, Census of Iowa, 1885, Manuscript Population Schedules for Harrison and East Des Moines Townships, Mahaska County.
Note: The total population was 1200.

assortment of foreign-born residents (Table 1.1). Muchakinock's population in that year totaled approximately 1,200. Ten years later, the state census for 1895 reported that Muchakinock contained a total of 3,844 people.[57] Of that number, 2,802 were white and 1,036 were black. Table 1.2 lists birthplaces of foreign-born residents of Muchakinock in 1895.

In 1885 and 1895, Swedes constituted Muchy's largest nonblack ethnic group. By 1895 they comprised 32.5 percent of the foreign-born population and 4.6 percent of the total population. The first

Table 1.2. Birthplace of Foreign-Born Residents, Muchakinock, 1895

Birthplace	Number	Percent of Total Population
Sweden	178	4.6
England	110	2.9
Wales	79	2.0
Ireland	49	1.3
Scotland	37	1.0
Austria	30	0.8
Hungary	25	0.7
Germany	24	0.6
Norway	8	0.2
Canada	5	0.1
Bohemia	1	0.0
Holland	1	0.0
Russia	1	0.0
Total	548	14.2

Source: Iowa, Census of Iowa, 1895, Manuscript Population Schedules for Harrison and East Des Moines Townships, Mahaska County.
Note: The total population was 3844.

Swedes arrived in Muchakinock in the early 1870s. According to a history of the Augustana Lutheran church, the first Swedes were probably Mr. and Mrs. Johan Olson and Jakob Olson from Dalsland and Johan Dahn from Smoland. Jakob Olson worked for the local coal company as an engineer for a hundred dollars a month, while Johan Olson and Dahn worked as coal miners. The Swedish coal miners were competitive among themselves and sometimes had contests to see which miner could load out the most coal. One time Johan Olson and Dahn won the contest by blasting and loading out 9,000 bushels in one month, with each bushel weighing eighty pounds. The miners were paid one dollar per ton for loading coal. In the early years of Muchakinock, housing conditions were rather primitive, as the Swedish miners, the history noted, "had to live for quite a while in houses that had no floors."[58]

Sometime in the 1870s, Swedish families began to meet for church services, although they did not officially organize a church until 1882. Before 1882 the congregation was served by Pastor M. F. Hokanson from nearby Munterville and by a layman, E. Anderson. Pastor Hokanson baptized the first Swedish child born in Muchakinock, Dahn's daughter Dina. Pastor M. P. Odin from Ottumwa organized the Swedish Lutheran congregation in 1882. Chartered members included Carl Otto Carlson, who came to Muchy in 1882 and moved to Buxton in 1901; Carl Swenson, who came to Muchy in 1881 and moved to Minnesota in 1891; Mrs. Thea Dahn, who settled in the area in 1873 and moved to Rock Island in 1891; and Aron Peterson, who came to Muchy in 1880 and moved to Minnesota in 1890. The Congregation also contained several families from Holland.[59]

In 1882 the Muchy Swedish Lutheran congregation built their first church. The parishioners had some trouble with the building because, a church historian recorded, "due to the depressions that appeared in the ground caused by the working in the mines, it was in danger of falling over and the members of the congregation had a lot of work to do to prop it up and fill up under and around with dirt." Members later moved the church several hundred yards west of the original site. It was constructed of wood and all interior articles were handmade. In 1884 the church joined the Augustana Synod. During these years, Pastor Odin from Ottumwa served the Muchakinock Church.[60]

Table 1.2 also shows that Muchakinock attracted residents from other northern and western European countries, as well as immigrants from eastern Europe. With the exception of the large black population, Muchakinock accurately reflected the ethnic composition of most Iowa coal camps in the late nineteenth century. In 1895, Iowa still contained a sizeable number of miners who had emigrated from the British Isles. Some four decades earlier, large numbers of English, Welsh, and Scottish miners had started emigrating to the United States seeking higher pay, better working conditions, and shorter hours. Most of these men settled in Pennsylvania, Ohio, and Indiana. During the next several decades, some families moved westward into Illinois and Iowa. As a result, English, Welsh, and Scottish miners constituted a significant proportion of Iowa's coal mining population in the 1870s, 1880s, and 1890s. By the latter decade, Welsh and Scottish mining families had moved farther west, but many English miners remained in Iowa. Muchy also contained a sizeable number of whites born in the coal mining states of Pennsylvania, Ohio, Indiana, and Illinois. This group totaled 466, or 22 percent of the community's white population in 1895. In many instances, these men had been coal miners before their migration to Iowa.[61]

During the 1890s, however, Iowa's coal mining population underwent an important transition. Throughout that decade, the number of immigrants from eastern and southern Europe gradually increased, so that by 1900 these groups were the most visible among the coal industry's foreign-born populations. Table 1.2 shows the beginning of that transition with the sizeable number of immigrants from Austria and Hungary. Taken together, Austrian and Hungarian immigrants constituted the fourth largest foreign-born group in Muchy in 1895.

Although Muchakinock contained a wide variety of ethnic groups, race relations there seem to have been harmonious. Herman Brooks, a black miner, remembered that there was little, if any, racial conflict. Brooks stated: "There was Irish, Welsh, Swedes, Germans, Polish and Negroes and we didn't know what color was. We all danced together, we worked together, we ate together, we played ball together. . . . We never had anything that wasn't mixed." Brooks concluded: "We even went to church together, and we danced together and everything." Brooks recalled that there was also cooperation between blacks and whites when it came time to butcher. He stated that butchering started on Thanksgiving Day: "We would kill four or five of them great big hogs, great big, and all the neighbors would help

Herman Brooks as a teenager in Buxton. Brooks was born in Muchakinock in the latter 1880s and moved to Buxton with his parents in 1900. (*Courtesy of Herman Brooks*)

one another. You kill one day and your neighbor would kill the next day and all the neighbors would help, and those Swedish people, they really knew what to do with a hog."[62]

Aside from limited personal remembrances, however, there is little evidence to show that blacks and whites often interacted on a personal basis. Certainly black and white miners worked together, and all frequented the same businesses, but lodges and churches were segregated. It is difficult, perhaps impossible, to determine whether Muchakinock itself was divided into ethnic enclaves. Census data indicate that the homes of some black families were clustered together, but there were also many areas where the homes of black and white families were intermingled. Interviews indicate that the Swedish miners lived by themselves in one area of town and attended their own churches. It is not clear whether other white ethnic groups clustered together.

There does seem to have been cooperation between blacks and whites in community affairs. Both races not only attended the town's major social functions but also cooperated in planning the events.

Muchy residents also worked with residents of other coal camps. School programs were sometimes presented in cooperation with schools at Givin, a nearby coal camp that was predominantly white.

The *Solicitor* and the *Bystander* included occasional accounts of violence in Muchakinock, but most of these do not seem to have been race related. On one occasion, however, George Taylor, the editor of the *Solicitor* and a resident of Oskaloosa, was beaten by six white men. Taylor and his wife were riding to the Muchy fairgrounds in a local hack when the attack occurred. The *Solicitor* reported that Taylor had charged that three of his attackers were "prominent citizens and businessmen of Albia and one of them is said to be a member of the Albia city council." It is not clear if action was taken against the men.[63]

The takeover of the Consolidation Coal Company in 1881 by the Chicago and North Western Railroad proved a propitious move for Muchakinock. From that point on, the camp developed rapidly and just as quickly became identified as a black community. Throughout the 1880s and 1890s, Muchy retained its reputation as a black community even though blacks became a minority. Blacks and whites alike apparently enjoyed a good standard of living and viewed the community as a good place to live. Outsiders also viewed Muchakinock in a positive light. In 1897 *Bystander* editor J. L. Thompson referred to Muchakinock as "the colored Athens of Iowa" and commented that it was a great pleasure "to see most all the people drive out in their own beautiful buggies."[64] Moreover, the black population seemed to be stable. The same names appeared again and again in the *Bystander* and the *Solicitor* throughout the 1890s. Many of these same names carried over to the Buxton news column after 1900.

Although Muchakinock contained many whites as well, the blacks formed a strong, close-knit community with highly visible leaders. Within that leadership, both males and females played active roles. Many black women participated in church, school, and lodge meetings, presenting papers on religious and educational topics and serving as delegates to church and lodge conventions. Blacks encouraged their people to improve themselves in every way; this theme permeated all their religious, civic, and political activities.

Within the community and the county, race relations seemed to be harmonious. The Swedes formed an independent community, and other white ethnic groups may have done the same, but there does not seem to have been outward friction between blacks and whites. Muchy's black residents frequently traveled to Oskaloosa, the county seat of Mahaska County, to patronize local doctors and to attend social functions, and black residents of Oskaloosa, in turn, frequently visited Muchy for church and social affairs.

When placed within the context of black history in Iowa, the black experience in Muchakinock seems highly unusual. Iowa had granted black males the right to vote only thirteen years before blacks arrived in Muchy, and the Iowa General Assembly had passed legislation that allowed black males to serve in the state legislature only one year before blacks arrived in the camp. Yet within the next decade Muchakinock blacks were running for public office and playing an active role in the state's Republican party. Perhaps it is the local economy that provides the key to the difference. Throughout the 1880s and 1890s, Consolidation's mines worked regularly. In the mid-1890s the Iowa coal industry experienced some difficult years, with miners facing unemployment, but this situation apparently did not affect Muchakinock as strongly as other Iowa coal communities. Miners in Muchy worked on a fairly regular basis throughout the 1890s, probably because of the community's connection with the Chicago and North Western. The fact that blacks continued to migrate from Virginia throughout the nineties indicates that the company regularly added new workers rather than dismissing old ones. Because few miners in Muchakinock were laid off as a result of the depression in the midnineties, the different ethnic groups did not have to compete for employment. Once basic economic needs were met, whites and blacks could live side by side comfortably, interacting on the public level but maintaining some private separation. These favorable economic and social conditions would continue into and beyond the period in the late 1890s when the Consolidation Coal Company, finding its coal reserves in Mahaska County nearly depleted, decided to move south into Monroe County and to replace Muchakinock with the new camp of Buxton.

2

THE CREATION OF A COMMUNITY

IN September 1903 the *Bystander* presented the "Buxton Souvenir Edition" to inform its readers of the wonders of a most unusual coal camp. The beautiful town of Buxton, the newspaper proclaimed, was "the model mining camp in Iowa, and perhaps in the United States." The community heralded by the *Bystander* had come into existence only three years earlier, when officials of the Consolidation Coal Company decided to relocate their company's operation from Muchakinock to Buxton. The move was completed in the summer of 1901. By 1905, Buxton contained approximately 5,000 people, 2,700 of whom were black.[1] From 1900 to 1910, Buxton flourished as Iowa's largest coal mining community. Although Buxton shared many characteristics with its predecessor, Muchakinock, the new community attracted a larger population and enjoyed a far greater visibility as a black coal mining community. For the first ten years especially, Buxton was known throughout much of the nation as an exceptional place for blacks to live and work.

In 1896 two important events took place that would eventually lead to Buxton's inception, as well as to its unusual character. In that year, Consolidation officials determined that the coalfields around Muchakinock would soon be worked out. The company hired Henry Newton, a mining engineer living in nearby Hamilton, to locate new coal reserves. After four years of prospecting, the company purchased 8,600 acres in Monroe County and 1,600 acres in Mahaska County at a total cost of $275,000.[2] The company then laid out the new camp of Buxton. The second important event in 1896 was the retirement of John Emory Buxton as general superintendent of Consolidation and the appointment of his son, Ben C. Buxton, as his replacement. The younger Buxton supervised the company's purchase of the additional land and its move to Buxton. According to the *Bystander*, the younger Buxton "designed and superintended the laying out of the town, the plans and construction of the buildings, the location and equipment of the mines, the water supply, the drainage, and all the interesting details in the development [of the town]." At the time he became superintendent, Ben Buxton was twenty-five years old, and he would retain the superintendency during the community's formative years, resigning in 1909.[3]

In June 1900, Consolidation began the process of creating a new community. The site selected was section 4 of Bluff Creek Township, located in extreme northern Monroe County. The county road dividing Monroe County to the south and Mahaska County to the north marked the northern boundary of Bluff Creek Township and the northern border of Buxton proper. In locating the Buxton camp, Consolidation officials apparently paid close attention to selecting an area with good drainage. Clayborne Carter recalled that many houses were located on high ground which was not the case in all coal camps.[4] Carter's views were echoed by a state sanitary engineer who surveyed the Buxton camp in 1919:

> From the standpoint of topography, Buxton Camp is nicely located. The site is rolling so that good drainage is provided. Practically all the houses are on the sidehills or the uplands, those on the lowlands being batch houses or small houses.[5]

The company immediately extended a branch of the Chicago and North Western Railroad from Muchakinock to the new field and started construction of Mine No. 10. They also began construction of miners' homes, and in October the first homes were ready for occupancy. Consolidation continued to build homes, and as groups of homes were completed, additional families made the move from Muchy. Apparently the company began building homes on First Street and then continued the construction from First Street to Second Street and so on until they had finished the home building.[6]

Consolidation gave each family several days' notice, and on the appointed day company employees arrived with horse-drawn wagons to haul furniture, personal belongings, animals, and poultry to the Muchy depot. Workmen then deposited the belongings on flatcars for the trip to Buxton. Herman Brooks related that when people moved from Muchakinock, "everybody didn't move at one time . . . because they couldn't build the houses fast enough. They [had] to get so many houses built, . . . and then you would get your orders to move." Bessie Lewis, who was thirteen at the time, remembered that in August 1901 her family, the Albert Rhodes family, moved with a group of a hundred families. When they arrived in Buxton, the Rhodes family was assigned the house numbered "12" on East First Street. The company continued to build houses through 1903. In March of that year, the *Bystander* reported that the company was completing another 375 homes.[7]

In planning the community, company officials decided to place four houses on each acre, alloting a quarter-acre for each home. The foundations of the homes varied. Some were wooden blocks or posts, while others were stone or rock. Workmen placed these materials in each corner of the house and along the sidewalls. Some families later boarded in around the foundation. None of the company houses had basements. All company homes were one and one-half stories, with either five or six rooms. The homes had two bedrooms upstairs and either three or four rooms downstairs. The main downstairs rooms were a kitchen, a dining room, and a front parlor. Six-room houses included a bedroom downstairs. The stairway was placed in the middle of the house, and on the second floor, doorways went off in opposite directions to the two bedrooms. All homes were plastered inside and had attractive wooden floors. Many families added summer kitchens on the back of their houses, which freed an additional room in the main house for other uses. Some families built front porches, and Lester Beaman remembered that some homes also had small back porches. The homes were originally painted red, but company employees later painted them steel grey. Miners paid seven dollars per month for five-room houses and eight dollars for six-room houses. It is not known precisely how many homes were located in Buxton, although the number of company houses was estimated at a thousand. According to a 1919 plat map of Buxton, the company had laid out fifty-odd blocks in the town. It is not clear, however, if homes were built in all fifty blocks.[8]

Because the company had created quarter-acre lots, each family had room for a large garden and several outbuildings. Each back yard contained an outhouse, a small coal shed, and a cistern. The coal shed stood close to the alley so that delivery wagons could unload the coal without coming into the yard. Each house had gutters and drainpipes to catch rainwater for the cistern. While some families planted trees

View of Buxton's residential area showing company houses with summer kitchens added on the rear, fencing around lots, outbuildings, and gardens. (*Courtesy of Iowa State Historical Department*)

and flowers in the front, several former Buxton residents recalled that their mothers frequently swept the back yards and did not allow any grass to grow there. Dorothy Collier remembered that her family "had to sweep their back yard every day . . . and pull up the little sprigs." Her mother planted some flowers in the front yard. Vaeletta Fields recalled that her mother, Minnie B. London, just "let the grass grow, . . . the weeds and everything." In addition to cisterns, at least some parts of Buxton had wells placed in the middle of the streets. Charles Lenger remembered that about one-half block from his parents' home there was a well that provided water for four or five neighboring families.[9]

The interiors of the company homes varied from household to household, but all informants described their homes as comfortable. Most families had two stoves in their homes, a coal cookstove in the kitchen and a potbellied heating stove in the dining room. The cookstoves had warming ovens above and water reservoirs on the side. Jacob Brown remembered that because cookstoves heated up the houses in the summertime, he had purchased a three-burner oilstove for the kitchen for summer use. Many kitchens also contained one-unit cupboards that included a flour bin and a bread box. Families often kept perishable foods in a bucket in the cistern, but some informants remembered that their parents had wooden iceboxes in which a block of ice placed in a top section cooled the lower part of the box. The kitchens were large, and several informants stated that most family activities took place there. Jacob Brown recalled that in the dining room many families had tables, chairs, and either buffets or china closets in which to keep their fine china. None of the company homes was electrified (the families used kerosene lamps to provide light) and none had indoor plumbing. Brown recalled that most families had linoleum on their floors.[10]

Some former residents remembered that their homes were furnished with fine-quality furniture. Marjorie Brown recalled that her parents' parlor had a carpet and a piano. Dorothy Collier related that her mother, Alice Neal, had a new cookstove in the kitchen. The Neal living room contained a green plush sofa, with arms decorated with carvings of a lion's mouth, and a Morris chair, a forerunner of the modern recliner. In the Neal home, the living room was off-limits except on Sunday. Then Mrs. Neal would "take a shovelful of coals, hot coals, and start a fire in the living room [stove], and we'd be in there on Sunday." The Neal family also had a stove in the children's upstairs bedroom.[11]

Informants stated that families who lived at the end of a street or had a ditch adjacent to their lot had bigger lots, so they could raise a few chickens and possibly a cow or pig. Herman Brooks' parents, for example, lived at the end of Seventh Street and they were able to keep a horse and a cow. Mike Onder related that for families who lived along ditches, especially, it was easier to keep animals, particularly hogs. Because they lived at the end of a street, the Kocur family was able to keep four or five hogs.[12] Apparently the company did not allow families to keep chickens or livestock on regular-sized town lots.

Almost all informants recalled that their parents had raised large quantities of vegetables and planted some fruit trees. Hucey Hart's parents "raised practically everything." Hart explained that in Buxton "a lot of them had their cows out there in the yard, chickens, hogs, and things like that. We had chickens and we raised cabbage and tomatoes and potatoes and . . . sweet potatoes and corn." Lester Beaman remembered that his mother raised many different vegetables, including collard plants and turnips. Hazel Stapleton related that her parents had "a lovely strawberry patch, and we had lovely gooseberry bushes and blackberries and raspberries. Then we had a few plum trees." The Stapletons also raised apple and cherry trees.[13]

Raising fruit and vegetables enabled most Buxton women to can large quantities of food. Bessie Lewis remembered that her mother, Margaret Rhodes, canned a great deal and made certain that her daughters learned the art. Susie Robinson recalled that as a young wife in Buxton, "I canned a lot, and after I got married I canned a lot of vegetables and I made big gardens myself. I made all the gardens by myself, great big gardens. . . . I loved to garden." Some families canned meat, particularly beef and chicken. Many informants recalled their parents butchering hogs. Sometimes this was a neighborhood affair, but often the family butchered alone. The pork was either fried down or salted, and sometimes smoked.[14]

The residents' practice of raising large gardens and keeping livestock and poultry underscores the self-sufficiency of most Buxton families, an important economic fact. In addition to providing most of their own fruit and vegetables, women also provided other foods for their families. By keeping cows, many women not only supplied their families with fresh milk daily but were also able to churn butter; by feeding chickens, they provided their family with both meat and eggs. Although all families purchased some food at the company store, often they found it necessary to buy only staples. Thus domestic food production not only aided them financially but also made them less reliant on the company store. In addition, these families probably enjoyed better and more flavorful food and maintained higher nutritional standards.

Not all Buxton families, however, lived in company houses. Numerous black families either leased or purchased small acreages nearby and built their own homes. Consolidation established the policy in 1900 that any employee could lease land from the company in parcels varying from one acre to twenty acres. The employee then built or moved in his own house. These larger acreages allowed families to raise substantially more vegetables, particularly potatoes, which they then sold in Buxton. Gertrude Stokes's parents leased twenty acres on the outskirts of Buxton on which her father, Minor Waites, did some farming. Waites then purchased a house from the nearby coal camp of

Lost Creek and moved it to their acreage. The Waites family raised a large garden that included many cabbage plants. In the fall, Gertrude's father dug small caves in the side of the hill to store the cabbages and throughout the winter her mother sold the cabbages to townspeople. In the summer Gertrude's parents took her to town along with a wagonload of fresh vegetables. Gertrude recalled that townspeople would "come to the wagon and buy what they wanted—green beans and peas and different things that we had."[15]

White families in Buxton also leased land in adjoining areas. A large number of Swedish miners moved their houses from Muchakinock to Buxton. Alex and Agnes Erickson remembered that many miners, including their father, had disassembled their homes in Muchy and loaded them onto railroad flatcars for transportation to the new camp. Once in Buxton, the miners unloaded the sections and put the houses back together. The company not only paid for the transportation of the houses but also rented each miner one acre of ground for fifty cents per month. The result was that homes owned by Swedes in Buxton often differed in size and design from the company homes.[16]

Some white families also purchased land in the area. Harvey Lewis related that his family moved to Buxton in 1913, when his father went to work for Consolidation as a coal miner. He recalled that his father "made good money in the coal mines and he bought a farm." The eighty-acre farm was located two and one-half miles south of Buxton. It is not clear if the older Lewis continued to work in the mines. Archie Harris, a lifelong resident of the Buxton area, followed the same practice. Harris worked for Consolidation for many years, but from the time he married, he and his wife had their own farm, located south of Buxton proper. They maintained considerable economic independence, as they seldom purchased anything at the company store.[17]

At the same time that Consolidation constructed company homes, it also built homes for its managers. The *Bystander* reported in November 1901 that Superintendent Ben C. Buxton's residence was almost completed. The home was situated on a hill a short distance west of Buxton's main business district. Ada Morgan, daughter of E. M. Baysoar, who worked as the general superintendent for Consolidation from 1909 to 1913, lived in this home for four years. Marvin Franzen's father was the caretaker of the house for the Buxton family.

Home built for Consolidation's general superintendent and first occupied by the Ben C. Buxton family. Photo taken as the home appeared around 1910. (*Courtesy of Ada Baysoar Morgan*)

Both Morgan and Franzen reported that the two-story house had four rooms and a bath downstairs, four bedrooms and a bath upstairs, and a full basement. The first floor contained a parlor, a dining room, a kitchen, and an office. The floors were hardwood throughout the house, with carpets and runners placed in and between the rooms. The house had central steam heat and electricity. The house also had both a front and back stairway. Franzen related that a screened porch was attached to the front of the house, which provided a commanding view of Buxton. A high limestone wall ran along the south and east sides of the house.[18] About a year after the superintendent's home was finished, the *Bystander* noted that W. H. Wells, the manager of the Monroe Mercantile Company (the company store), had moved into his new residence. The home was equipped with electric lights and steam heat. The company also built five or six homes for other company officials on First Street. These homes were larger than the miners' company homes and were electrified.[19]

As more families moved into Buxton, it became apparent that the community would include not only the land owned and maintained by the company but also several outlying sections leased from the company or privately owned. In June 1901 the *Bystander* reported that workmen had completed the foundation for B. F. Cooper's drugstore, located just north of Buxton proper in a suburb called Coopertown. Cooper had purchased several lots from Reuben Gaines, Sr., in Mahaska County just north of the Monroe County line. Eventually Coopertown also included the J. W. Neeley grocery store, a dance hall, and several private homes. The area soon developed a tough reputation because of several clubrooms located above Cooper's drugstore where local residents could play cards and order alcoholic drinks.[20] Leroy Wright remembered Coopertown:

> Nothing around there but people bootlegging and having beer and everything like that, you know. And they had a little place there, they had pool tables there where you go and play pool and where they bet for money and all like that. . . . If things [got] too bad and [the owners] couldn't do nothing, this constable would call the sheriff from Albia.[21]

Coopertown was only the first of many small outlying areas or suburbs that eventually encircled Buxton proper. Generally, references made to Buxton by contemporaries or later by former residents include both Buxton proper and the outlying suburbs.

Consolidation's policy of leasing land led to the creation of two additional suburbs, East Swede Town and West Swede Town, which were located east and west of Buxton proper. Most Swedes who moved from Muchakinock, where they owned their own homes, decided to take advantage of the company's leasing offer. According to John Baxter, East Swede Town "was a town in itself" with a school, a blacksmith shop, Emil Larson's grocery, and a Swedish Lutheran church known as Ebenezer Lutheran. Mike Onder also remembered that East Swede Town was "just like a little town" of two streets with houses on both sides. East Swede Town also contained an elementary school, known as the Swede Town School, that most Swedish youngsters attended. West Swede Town contained two churches—the Swedish Methodist and the Slovakian Lutheran—but it had fewer homes than East Swede Town.[22]

Other outlying areas or suburbs included Gainesville, Hayes Town, and Wells Town, or Wells Hill. Gainesville was founded by Reuben Gaines, Sr. Gaines had lived in Muchakinock, and in the late

1890s, when Consolidation began discussing the possibility of relocating in Monroe County, he invested in land in the general area of the proposed camp. He purchased two forty-acre plots in Mahaska County for fifty dollars an acre. One plot, located directly across the county line from Buxton proper, came to be known as Gainesville. Although Gainesville did not have a post office, according to Reuben Gaines, Jr., it was listed on the Iowa map around 1915. Gaines, Sr., built several businesses, including a barber shop and a hotel. Eventually Gainesville also included a livery stable, the Hartman Dye Works, a meat market, a shoe store, and approximately eight to ten houses. Hayes Town, located in Mahaska County near the northwest corner of Buxton proper, contained homes owned by W. L. Hayes and included a meat market, a drugstore, a photograph gallery, a livery stable (including a hack line), and a millinery shop. Wells Town was a collection of about a dozen houses built by W. H. Wells. It was located southwest of Buxton proper and south of West Swede Town. Wells owned a home in the area. According to Lara Wardelin, whose family lived there in the early 1900s, the houses and acreages were larger than those in Buxton proper.[23]

In addition to being surrounded with outlying suburbs, Buxton proper also contained two distinct sections known as Gobblers Knob and Sharp End. Sharp End was located in south Buxton and lay south of Fourteenth Street, while Gobblers Knob, designated by some as the most fashionable part of town, was located in west Buxton proper. It is not clear how these areas got their names, but many informants remembered them as distinct parts of town. Gobblers Knob was apparently one small area of the residential district, but Sharp End was a larger area that contained not only homes but numerous businesses, two churches (a Seventh Day Adventist church and the Tabernacle Baptist Church), and an elementary school.[24]

Sharp End had a substantial business district that was separated by approximately twelve blocks from the Buxton's main business district. The area included John Baxter's meat market, a restaurant, the Stryder Grocery, a barber shop, the Lucas Bakery, the Hutchison Drug Store, the Jack Brookins Tailor Shop, the Blue Front Dance Hall, Manny Lobbin's Livery Service, and the Williams Store. Several store proprietors and their families lived above their shops. The Lucas Bakery sold bakery goods to several restaurants in town, delivered bakery goods to nearby coal camps, and operated a delivery wagon in Buxton. Mike Onder remembered that as a young boy he could go into the bakery and buy a large sack of broken cookie pieces for five cents. In his livery stable, Manny Lobbins provided recreation for young people, as he kept a few ponies that he rented out for rides. Earl Smith recalled riding the ponies as a young boy: "We thought that was great, riding a horse ourselves. John Baxter used to give me and his nephew money to go out there and get one of the ponies."[25]

The most prominent feature of Sharp End, however, was its tough reputation. Many former residents recalled that several places in Sharp End sold liquor and provided local people with a place to gamble. Mike Onder stated that "somebody got killed there about every weekend. . . . By golly, every other house there, they sold that chalk," a drink described as tasting like wine. Bessie Lewis remembered that "They had everything it takes for gambling. Even when it comes to fighting and fussing and killing one another, they done that too." Lola Reeves, a former schoolteacher in Buxton, remembered hearing that Sharp End was a rough part of town. She concluded: "They used to talk about a lot of trouble out at Sharp End."[26]

The presence of some six suburbs around Buxton suggests that even though company officials had laid out the original town in careful fashion the gradual addition of private businesses and homes eventually gave the town a rather random appearance. In 1910, for example, a contingent of Iowa newspaper editors meeting in Albia traveled to Buxton, and one editor, Stanley Miller of Mt. Pleasant, recorded his impressions, commenting that he had long heard about Buxton and had eagerly awaited the trip. When he arrived, however, he found that there were few principal streets in the town. He added: "There are a good many business houses in the town, but there is no business section. You will find a grocery store, or a furniture store nearly anywhere, and it is usually surrounded by residences."[27]

Residents were able to locate businesses anywhere because Buxton was not incorporated and therefore lacked zoning ordinances. Without incorporation, Buxton also had no municipal government, no significant law enforcement structure, and no municipal maintenance staff or equipment to provide upkeep for roads and streets. The company apparently maintained a few major streets within the town, but many went unattended. Buxton contained no sidewalks in the residential areas, but company officials or homeowners put down cinders, furnished by the power plant, to form paths and roads. The company hired two security officers to look after the safety of company property, and apparently Monroe County officials assigned one deputy

View of Buxton's East First Street from the corner of Main and First streets before 1912. The first company store is the white, two-story building on the left, and the large YMCA is the three-story structure on the right. (*Courtesy of Iowa State Historical Department*)

sheriff to Buxton. Consolidation maintained its own property, such as the company store, while individual property owners managed theirs.[28]

While Buxton's suburbs and residential areas were taking shape, the town's main business district was also under construction. Located in north Buxton on First Street close to the Chicago and North Western Railroad tracks, the main business section contained the company store and approximately twenty other businesses. Built in 1901, the first company store (it was destroyed by fire and replaced in 1911) dominated the main business district. The store was originally known as the W. H. Wells Company Store and later as the Monroe Mercantile Company. Store management personnel included R. R. MacRae, general manager; H. B. Henderson, secretary and treasurer; and E. M. Hammond, purchasing agent. Hammond maintained his headquarters in Chicago but also spent considerable time in New York and St. Louis. The store management believed that this method of buying made them competitive with other stores in the area, as well as allowing them to display the latest styles and fashions in clothing, furniture, and dry goods.[29]

The company store was divided into departments, with each department having its own manager and conducting its business independently of the others. The individual departments included shoes, shoe repair, clothing, hardware, furniture, and groceries, in addition to a pharmacy and a hay and grain section. Some departments also provided other repair services. Bessie Lewis remembered, for example, that W. H. London, the manager of the hardware department, repaired cookstoves. The store also put up its own ice—about 3,000 tons a year—and two wagons delivered ice to consumers during the summer. A soda fountain provided a gathering place for younger Buxton residents; in 1907 the *Bystander* noted that "the company has its own ice cream factory, the freezer and ice crusher being run by electric motors." The company store sold phonographs and records on the second floor, and Clayborne Carter recalled: "Here the boys would loaf and have the girls play the records." The store carried name brands, including Red Cross shoes, Stetson hats, Kuppenheimer suits, Heinz bottle goods, and Armour canned meats. If the store did not carry an item a customer wanted, it could be ordered, whether it was an automobile, a tailored suit, or a piece of furniture. In 1907 the *Bystander* editor commented: "It seems that one of Chicago's department stores was removed to this new colored mining camp."[30]

The store also contained a complete undertaking department. It owned a hearse, a landau (a two-seat carriage with a folding top) and several plain carriages. The *Bystander* reported in December 1907 that "full charge is taken of funerals even to the preaching of the burial service, if necessary. Mr. Sam Welch, as undertaker and funeral director, has been with the company a number of years and he fully understands his work."[31]

Because of the large number of shoppers at the Monroe Mercantile Company, store personnel devised a special ordering system whereby customers' ordering days were spread throughout a two-week period. Carl Kietzman recalled that the company assigned ordering days to families on the basis of their residence; all residents of a particular street ordered on the same day. Gertrude Stokes's mother went to the store on the first and sixteenth day of each month to place her grocery order. Residents also had the option of remaining at home and having clerks take their orders. Clerks boxed up the orders, and eight men delivered the groceries to the various residences. If families needed additional items between ordering days, they went to the store and purchased the items outright.[32]

The company store also served as a social gathering place for many local miners. The *Bystander* noted that the store at night, after the miners had eaten supper, "is one large auditorium where they gather to trade, smoke and talk. No comment is made on loafing—it is far better to see them thus passing their time than in saloons just outside the town. But they are orderly. Boisterous or blasphemous language is not permitted."[33]

The company store hired both black and white employees, often totaling as many as seventy or eighty people. Employees included several general managers, department managers, clerks, cashiers, undertakers, and employees involved in the production of ice cream and ice. The *Bystander* noted that out of seventy employees about fifteen were black. The editor added that "Mr. MacRae [the store manager] is always glad to get competent help, but there are too few of our people who realize the importance of preparation of education for such work. A refined, well educated Negro has always been in demand by this company."[34]

Opinions of former residents varied as to whether prices at the company store were higher than those in private stores. The *Bystander* editor emphasized that although the company store was a monopoly, "it is a monopoly that is hard to classify. Merchandise is of better quality and sold at a lower price than most stores in the state." He added: "Prices are extremely moderate, comparing favorably with many of our Des Moines stores. In fact, many articles are sold lower than even in our city where competition is so keen." Although many former Buxton residents did not recall that company store prices were high, some did. Alex Erickson believed that prices at the company store were higher than at other stores because the store extended credit. He remembered that some miners would depart in the middle of the night, leaving large debts at the store. To compensate for the unpaid bills and for the extension of credit, Erickson believed that it was necessary for the store to charge slightly higher prices. Mike Onder also remembered that the company store prices were a little higher, perhaps a nickel higher on most items. When his mother started trading at other stores, she was able to make her purchases a little cheaper. Like Alex Erickson, Onder believed that the store was justified in charging slightly more.[35]

While most Iowa coal operators required their employees to trade at the company store, Consolidation followed a more lenient policy. Almost all informants stressed that the company did not insist that they trade at the company store, although all stated that they did most of their trading there, because of its convenience and wide selection. Carl Kietzman remembered that the company sometimes used indirect pressure to convince its employees to trade at the Monroe Mercantile Company. Kietzman recalled: "Well, they didn't really put the pressure on, but if you didn't buy at the company store . . . then maybe you'd have to wait a long while for the room. . . . They wanted you to trade at the company store but you made your money and you could spend it wherever you wanted to."[36] Kietzman was referring to the fact that after a miner had worked out his mining room to a particular size, he then had to wait for assignment to another room. At that point the foreman could show favoritism toward certain miners by placing their names at the top of the list, or conversely show disfavor by putting miners' names at the end of the list.

Unlike most company stores, which catered only to camp residents, the Monroe Mercantile Company actively sought the patronage of residents from surrounding areas as well. In 1907 the *Bystander* editor noted that "the store is a purely business enterprise and goes after trade of the town and surrounding country, using common sense business methods." The editor commented that many people from towns of larger populations "within a radius of thirty miles come to Buxton to make their selections such as could not be supplied by their home merchants." The company store also provided both Buxton and area residents with a market for fresh produce and eggs. Many former residents remembered that their parents sold fresh produce, such as potatoes and tomatoes, to the company store, and rural residents brought in eggs and cream on a regular basis. Sister Maurine Sofranko remembered that her grandfather, who farmed several miles outside of Buxton, exchanged butter and eggs for groceries on a regular basis. The *Bystander* noted: "Farmers for fifteen miles outside of Buxton are regular customers. And why? They are paid more for their produce. They get better prices."[37]

Customers made purchases at the company store through a checkoff system, which was a form of credit. Gertrude Stokes recalled that when she went into the company store to shop, she first went to the business office, where she was issued a check or a card with numbers—such as $1, $2, or $5—printed on it. As she went to the different departments, the clerks punched out or clipped off parts of the check that equalled the amount of her purchase. Before the next payday, company bookkeepers subtracted the full amount of the card or check from the amount that Gertrude Stokes's father earned as a miner. Hobe Armstrong's meat market was the only other business in Buxton where residents could use the company checkoff system.[38]

The checkoff system enabled company employees to purchase almost anything they wished, and there were few, if any, limits on the total purchases. Some former residents remembered that on payday all they had was a "snake." This referred to a red line drawn through their wage statement. If the miner and his family had spent all that the miner had made in the previous two weeks, the red line indicated that the miner received no pay on payday. In some cases, the total came to more than their wages. In that event, bookkeepers subtracted the additional amount from the following paycheck. Robert Wheels remembered that there was no question but that the company got their money. If an employee owed the store a hundred dollars, "you didn't get nothing until it was all paid off." And, he added, "if you ever once fell behind, you were behind for a long time."[39]

The company store also contained the Buxton Savings Bank, but it is not clear where the bank was located in the store. In November 1905 the *Bystander* stated that "a general banking business is done, drafts issued, deposits received and collections made." The *Bystander* noted that the bank was under the management of the Monroe Mercantile Company and that it served an important function for the residents: "The bank stands as a great promoter of thrift and frugality among the miners of the vicinity. The deposits are steadily on the increase showing a greater tendency on the part of the colored people to save."[40]

In 1911, Buxton experienced several fires. On February 21, Buxton's worst fire destroyed the company store and all its contents. Company officials converted the main YMCA into a store, however, and soon it was "business as usual." Consolidation announced plans to rebuild immediately. Less than a month after the fire, store manager R. R. MacRae had blueprints ready for the new company store. The structure was to be one story and would measure 162 by 120 feet. Sliding steel doors would be placed between each department to provide fire protection, and skylights would help provide interior light. By April 21 construction of the second store was underway.[41]

The second company store on opening day in 1911. In the foreground is the partially constructed White House Hotel. The mine superintendent's home is seen in the background between the two smokestacks. (*Courtesy of Dorothy Collier*)

Six months later the new company store was ready for business. The *Bystander* proudly noted that "There is no finer department store in the capital city," and that the company was continuing its policy of hiring black workers. Eighteen of the eighty-five employees were black, and the *Bystander* especially singled out E. C. Strong, a black man who was the company's longest-serving employee. Strong managed the teamsters and supervised all deliveries.[42]

Company officials eagerly pointed out the new store's modern features. The facility contained an elevator-escalator and a central bookkeeping system. With the latter, the store had a single cashier located in an elevated area in the rear of the store. Wires stretched from this office to various sales locations on the main floor. A clerk placed money in a cylinder-shaped container and, by pulling a chain, propelled the container to the cashier's office. The cashier recorded the sale, made change, and returned the container to the sales floor. Odessa Booker recalled that as a clerk in the store she was so short she had to stand on a box to reach the chain. The store also contained the Buxton Savings Bank, operated by the Consolidation Coal Company. The bank was located in the northeast section of the store and faced along First Street.[43]

All former residents interviewed remembered the second company store as an impressive shopping facility. Susie Robinson exclaimed: "The company store had everything you could mention. . . . They had a hat store in there and they had funeral homes in there and they had whatever you name. . . . They had it right in that company store. It was one block long." Agnes Erickson, who worked at the store as both clerk and bookkeeper, remembered that the store purchased everything in carload lots. The railroad tracks ran behind the store, so a chute was devised whereby the goods from the cars could be unloaded directly into the store's basement. Bessie Lewis recalled that if the company store did not stock an item they could order it: "Just say for instance that you wanted to get some furniture. Well, there wasn't any furniture there . . . and wherever you wanted to get this furniture—maybe you'd been to Albia or Eddyville or Belle Plaine, someplace where they had these stores—all you had to do was just let them know it and they'd send and get that stuff for you." Clayborne Carter remembered that the new store even had a window decorator from Chicago.[44]

Store policies remained much the same. Most miners traded there, but the company did not prohibit employees from trading elsewhere. Credit was extended almost without limit. The store continued to be well managed and drew high praise from former residents, who remember it as an outstanding place to shop. Moreover, the second store, like the first, continued to serve as the focal point of the town's main business district. Mike Onder recalled that the hardware department in particular provided a place where miners congregated after working hours.[45]

In addition to the company store and Armstrong's Meat Market, the main business district (located along First Street) included the two YMCAs, the Perkins and Son Hotel, the Thomas Brothers Drug Store, the J. H. Williams Drug Store, the D. L. Thomas Restaurant, Chicken John's Restaurant, and the Jeffers Brothers Restaurant. The main business area also included two additional restaurants, three barber shops, a post office, a depot, a lumberyard, a telephone office, the company doctors' office, several company barns, two churches, a dentist's office, and a pool hall. It is not known precisely where each business was situated, however. Located some distance back or northwest from First Street were stockyards, a soft-drink factory, a laundry, and several buildings owned by Consolidation, including its pay office, machine shop, heating plant, and warehouse.[46]

While the company store dominated the main business district,

Hobe Armstrong's meat market, located a short distance northeast of the company store, ranked second in importance. Armstrong raised some of his own beef, but also bought both beef and pork from local farmers. Armstrong had a slaughterhouse located south of his farm, and he also owned an ice house. Archie Allison remembered that Armstrong slaughtered his meat daily. Allison and his father frequently butchered for Armstrong. Allison recalled that he and his father would work all day in the mines, and in the evening "the old man would say we got to butcher tonight" because Armstrong needed the meat. Armstrong ran daily meat routes through Buxton and the surrounding coal camps. According to Allison, Armstrong used Indian ponies on the meat wagons, and the ponies were trained to stop every time a housewife came running out to place an order. Armstrong's son-in-law, John Baxter, had a meat market in Sharp End, and apparently Armstrong and Baxter worked closely together in purchasing and processing their meat.[47]

Lewis Reasby's lunch wagon, located in front of the main YMCA, provided another familiar sight in Buxton's business district. Reasby's son, Harold, described his father's place of business as resembling a small railroad car. The older Reasby worked in the mines during the week and operated the lunch wagon on weekends. Originally the lunch wagon was closed during the weekdays, but when Reasby's daughters became teenagers they operated the wagon during the week. About the same time, Lewis Reasby quit his miner's job and worked in the lunch wagon full-time. Reasby had saved money to purchase the lunch wagon by peddling food—wiener sandwiches and boiled eggs—to miners who on payday would head into the park or into the surrounding timber to gamble. Reasby managed to save $250 from his food sales and then approached a Mrs. Christian, an employee of the Buxton bank, who loaned Reasby an additional $250. The $500 covered the cost of the lunch wagon. Reasby sold hamburgers, fried chicken, fried chicken livers, ice cream, soft drinks, candy, and cigarettes. Mrs. Reasby fried the hamburgers and chickens at home, and on many weekends Reasby sold from 200 to 250 chickens. When he needed more chickens, he sent one of his daughters home to bring more down to the lunch wagon. Lewis Reasby also kept a gasoline stove in the wagon and sometimes fried beefsteak and potatoes for men coming home from the mines in the evening. The lunch wagon held about seven or eight men. In the summertime,

Reasby raised a variety of vegetables, particularly tomatoes, in a large garden and served these at his lunch wagon too. Mike Onder concluded that Reasby "had made a fortune" in his lunch wagon business.[48]

In 1903 a new structure appeared on First Street that was destined to play a major role in the social lives of Buxton's residents. In that year the Consolidation Coal Company contributed $20,000 for the construction of a YMCA. Plans called for the three-story building to measure 59 by 114 feet and to be located on First Street southwest of the company store. The first floor would contain rooms for YMCA officials, a secretary's office, a reading room, a game room, a library, a writing room, a kitchen, classrooms, a gymnasium, and a large bath department. The second-floor plans called for a large assembly hall (purported to seat between 800 and 1,000 people) and several smaller rooms. The third floor would be devoted to lodge activities, and the rents from these organizations would be used for building maintenance.[49]

Although the YMCA was eventually used for social events by both blacks and whites, the *Bystander* clearly indicated that it was originally intended as a facility for the black people of Buxton. The *Bystander* wrote that the building was to be "devoted to the uses and under the management of colored men." Minnie B. London remembered that the "Y" was built "expressly for the colored miners." She explained that when the black men "seemed reluctant to take advantage of the opportunity the superintendent indicated that he would turn it over to the white people. Our people, after reconsideration, pledged cooperation and then a very efficient secretary in the person of L. B. Johnson was engaged." In 1910, a visiting newspaper editor estimated that from 300 to 600 black youths used the facilities daily. The same editor noted that a white YMCA did exist, but that it was only a room located above the post office. Clayborne Carter recalled that the white "Y" did not survive long because it did not have a permanent secretary.[50]

Several years after the main YMCA building was completed, the company added a swimming pool to the facilities. The pool was housed in a small building adjacent to the main YMCA. The large room above the swimming pool served as a meeting room, and residents occasionally held dances there. The two buildings were connected with an enclosed walkway. In November 1910 the *Bystander*

Buxton's two YMCAs. The large Y had an auditorium, meeting rooms, and gymnastic area, while the small Y contained a swimming pool. (*Courtesy of Dorothy Collier*)

proclaimed "You ought to see that big swimming hole the boys of the "Y" have—and right in the house too, with a steam pipe in it so the water can be made warm for winter swimming."[51]

By 1910, Buxton had also acquired several community facilities. In 1902, Consolidation donated land in northwest Buxton for the creation of a park. Containing about five acres and known as the Buxton Island Park, the site served as a facility for numerous community functions. Local residents held the annual Fourth of July barbecue at the park, and traveling circuses set up tents there. Consolidation constructed a bandstand in the park that served as the location for Saturday-night band concerts presented by the Buxton Cornet Band. The park project was a community affair in that the Buxton Cornet Band helped develop the area. Consolidation also contributed land for a baseball park and two tennis courts.[52]

Throughout the first decade of its existence, many blacks operated businesses in Buxton. Hobe Armstrong's meat market was the largest black business, and B. F. Cooper's drugstore was also well known. Reuben Gaines, Sr., was a major real estate dealer and businessman in the Buxton area. These three men were described as the wealthiest black men in the county. George Woodson, a black lawyer, moved from Muchakinock to Buxton in 1901 and once again hung out his shingle. The community also contained a second black druggist, W. J. Watters. In 1907 the *Bystander* noted that James Roberts of Buxton was the only black cigar maker in Iowa and speculated that he might be the only black cigar maker in the entire Middle West. The two London brothers, A. E. and W. H., operated a music store and sold insurance. W. H. London also managed the hardware department at the company store. W. L. Perkins owned the Perkins Hotel, managed by his son, Lewis. Buxton contained five restaurants, two of which were operated by blacks, the Jeffer brothers. Two other black brothers, the Neelys, operated a grocery and dry goods store. The town also contained three black barbers. Two black women, Susie London and Laura Gaines, operated the town's only two millinery shops. Although Buxton's postmaster was a white man in 1907, he had three black assistants. Dr. Charles S. Taylor, a black physician, had a private medical practice.[53]

Buxton also contained several cooperative business ventures maintained by black men. In December 1907 the *Bystander* noted that the Buxton Laundry and Bakery Company "deserved much praise and mention" because the firm was "managed and controlled" by fifteen black miners who "are 15 thrifty, hard working men, working in the mines during the day and pushing their enterprise at night." The *Bystander* described the laundry as "the largest colored laundry in the middle west and possibly the entire west." In addition to a manager, the laundry hired six workers, all black. The same men also managed the bakery, described as having the capacity for 300 loaves at one baking. A second cooperative venture, the *Buxton Gazette,* was operated by a "company of colored men."[54]

The *Bystander* frequently commented on the dominant role played by Buxton's black population. In August 1902, *Bystander* editor J. L. Thompson, visited there. He subsequently reported that the company had recently built 125 new homes, adding that "[the blacks] have complete control of this town." As evidence of this, he noted that:

> the two constables are colored, the justice of the peace is colored, the only meat market is owned by Mr. H. A. Armstrong, who was postmaster at Muchakinock, the two restaurants are owned by colored people, the two barber shops are owned by colored men. The postmistress, Miss Anna Willis,

is one of the most highly respected young ladies of the town. The only band is the famous Buxton Colored Band of 31 pieces, the finest of its kind in Iowa, led by Professor A. B. Jackson, colored. There are five colored clerks in the large company store: Mr. W. H. London, A. B. Jackson, A. E. London, John T. Washington and Mr. Jones from Centerville. The largest drug store is owned by our good and genial friend, B. F. Cooper and he is doing a good business. The only saloon and billiard hall is owned by a colored man, Mr. Gaines. . . . The electrical engineer that runs the electric plant is a colored man, Mr. Abe Hart, and he knows his business. The two blacksmiths are colored. The company carpenter is colored, Mr. Ben Tate. There are several colored men owning valuable tracts of land and building a nice house.[55]

The diversity of Thompson's list of black businessmen indicates the substantial position they had attained by this early date.

By 1910, at the end of its first decade of existence, Buxton had become a thriving community with a population just under 5,000. Although it had ceased to be predominantly black by 1910, blacks still comprised the largest ethnic group, with 42 percent of the total population.[56] The Consolidation Coal Company remained the largest coal operator in the state, operating four mines and employing approximately a thousand people.[57] Iowa's coal mining industry, moreover, had remained prosperous throughout the first decade of the twentieth century. As a result, Buxton's coal miners and other company employees experienced full-time employment, and the town itself enjoyed considerable prosperity. By 1910 the community had reached its peak population.

From all reports in that year, Buxton seemed an ideal community destined to live forever. Not only were residents optimistic about the community's longevity, but they had every reason to think that the prosperous times would continue. Consolidation's employees made excellent wages, and that fact was reflected in a high standard of living. Moreover, Buxton's residents enjoyed excellent housing and shopping facilities, which was highly unusual for a coal mining community. In the minds of many residents, Consolidation deserved high praise for its creation and continued support of the town. Implicit in comments by residents, particularly blacks, was the attidude that the company had provided them with the good life and that Buxton's future was unlimited. At times, white miners also shared these views. Archie Harris recalled: "I didn't know Ben Buxton, because I was just a little kid [then], but he must have just loved the colored people. I guess he did, really, because he made everything so nice for them here in Buxton." Alex Erickson, reflecting on his life in Buxton, commented that although people worked hard in Buxton he didn't think that "anyone had a care there." He added: "You weren't afraid, you didn't worry, you didn't think about the next day. You figured the mines had been there and the mines would stay as long as you lived." Bessie Lewis echoed the same sentiment. Reflecting on the comfortable life-style that she, her family, and her friends had enjoyed, she mused: "In Buxton, you didn't have to want for nothing."[58]

3

WORKERS IN A COMPANY TOWN

HE Consolidation Coal Company constituted the central focus of life in Buxton. Consolidation operated the mines and controlled the town. As employer, Consolidation determined who would work, the type of work to be done, and the conditions under which the work would be performed. But the nature of the coal industry, which required coal operators to provide housing for their employees, and the subsequent appearance of company towns meant that Consolidation would also serve as general manager of the community. The story of Buxton as a company town must therefore include the study of three related but distinct areas: (1) the workers' various roles and the issues related to their work, (2) the company's role as an employer, and, (3) the company's role as a manager of community life. This chapter will deal with the role of workers in a company town, and the next will deal with Consolidation's wider role as employer and community manager.

Like most coal mining communities, Buxton contained one dominant industry. As a result, most of the town's male population went to work in one of Consolidation's mines, and coal mining presented male breadwinners with an occupation that was different from that which many of them had previously known. As an occupation, coal mining was a dirty and dangerous activity. Most men spent their working hours underground, often laboring under cramped conditions and frequently breathing air polluted with coal dust and sulphur fumes. Yet at the same time the men encountered some favorable working conditions. While coal miners found their lives aboveground to be closely supervised and often controlled by their employer, they found considerable freedom in their work belowground. Coal miners considered themselves somewhat independent workmen, and indeed they were. Once underground, miners went about their daily routines with little interruption or supervision.

In 1903, Consolidation operated three coal mines: No. 9, located five miles northeast of Buxton; No. 10, located approximately two miles southwest of Buxton; and No. 11, located about three miles south of Buxton. Because Consolidation's mines were all located at least a few miles from the community itself, the company provided train transportation for the workers. Each morning at 6 A.M. a train of eight coaches left Buxton carrying the miners to work, and every evening it transported the men back home.[1]

Once at the mines, the men boarded cages, or elevators, usually in groups of ten, to be lowered into the mines. This operation might take as long as forty-five minutes. Once underground, each miner, or digger, proceeded to his room. Upon entering the room, the miner was in what resembled a long tunnel. At the far end was the face, or coal seam; each day during the mining process a little more of the seam was removed. The average coal room measured thirty feet in width, and most rooms were mined out to a length of approximately 150 feet. Miners left the coal seams intact along either side of the room to provide support for the area being worked. The height of the room depended on the height (or width, as the miners called it) of the coal seam. In the Buxton mines, seams measured from three to six feet in height, which meant that many times the miners had to work hunched over or on their hands and knees.[2]

Miners sometimes worked alone in their rooms, and sometimes in pairs. Each day the miners received a certain number of coal cars, called a "turn." Typically, each miner could load out more cars than he received on any given day, so to increase his turn he might take a helper underground, usually a son, cousin, or nephew. By taking a helper, the miner could increase the number of coal cars assigned him for the day. This practice encouraged many men to take their sons underground before they had reached the legal age for underground work.[3] Sometimes two fully trained miners worked together in the same room and simply divided the total tonnage. Each man then received half the money earned.

The miner's daily work routine included a wide range of tasks. Entering his room in the morning, he first tapped the roof to determine whether the slate was tight or loose. If it sounded loose, he immediately placed wooden props under the suspect areas. During

View of an underground mining room after coal was blasted loose from the seam. The next step was for the miner to load chunks of coal into coal cars. Note the coal seam in the background and the slate roof. (*Courtesy of Iowa State Historical Department*)

the day the miner also had to determine where he would set his blasting shots so that coal could be loosened that night for the following day. He then drilled the holes in the coal seam, usually two or three, and filled these with blasting powder. Sometime during the shift, the shot firer came through the mine to inspect the holes. If they were not drilled correctly, the shot firer had the authority to condemn the shots and demand that the miner make new ones. During the day the miner also had to plan for the removal of a portion of the floor. He did this by drilling a hole in the floor for the placement of another shot. This practice, called "brushing bottom," lowered the floor in the room about twelve or fifteen inches and, in turn, lowered the railroad tracks. This practice provided the mule that pulled the coal cars with more headroom and allowed the miner to scoop coal into the cars more easily. Throughout the day, mule drivers came through the mines picking up loaded cars and delivering empty ones. By the end of his shift, the miner had propped his roof, loaded his assigned coal cars, and prepared his shots so that coal could be blasted loose for the following day.[4]

Although coal miners had to possess a considerable number of skills to carry out their work competently and safely, the craft as a whole did not require any formal training before men went underground. Most young men entering the mines for the first time were accompanying their fathers, uncles, or older brothers. Initially they ran errands, assisted in placing props, and helped with shoveling coal, but at the same time they observed the experienced miners' work. In this way they learned about the composition of the seams so they could correctly drill holes and tamp powder. After the newcomers had learned the basic skills, the foreman then assigned them their own rooms. Although the age varied at which young men became independent miners, many had reached that point by age eighteen.[5] Because coal companies hired men as miners without previous mining experience, the newcomers could move into the industry easily and begin to draw a paycheck immediately. Though they first worked as helpers, they were in effect being paid for learning the skills of the independent miner.

The fact that men could be hired without mining skills or previous mining experience held great attraction for both southern blacks and immigrants. Most of these men had little or no money and little education. Upon their arrival in Iowa, they needed to find employment quickly that provided an adequate livelihood for their families. Many men coming from either the South or Europe possessed some agricultural experience, but the move to Buxton gave them a point of entry into the industrial world. For young men born in Buxton (or Muchakinock), however, coal mining represented their only skill. Reflecting on his early life in Buxton, Robert Wheels commented: "I knew nothing else, you understand. I knew nothing else. Please remember that."[6]

Along with the actual miners, Consolidation also hired many other employees. Any person employed by the firm for work other than as a miner was known as a company man. In turn, company men fell into the categories of management or manual worker. Management personnel included the superintendent, the top foremen, and the general foremen. Company men who performed manual labor could be employed above- or belowground, but the great majority worked belowground. The company men performed a wide array of auxiliary services that insured that the loaded coal reached the top as quickly and efficiently as possible. In 1905, Consolidation employed 1,234 men (not including boys). Of that number, 94 percent were miners, as shown in Table 3.1. Ten years later the total number of Consolidation employees had dropped by 161, and miners comprised an even greater proportion of the total employees.[7]

Table 3.1. Percentage Distribution and Number of Workers by Occupational Group in Coal Production, Buxton, 1905 and 1915

Occupational Group	1905		1915	
Miner	94.0%	(1,160)	97.0%	(1,042)
Company personnel (management)	0.9	(11)	0.7	(7)
Company personnel (clerical)	0.2	(3)	0.2	(2)
Company personnel (manual, belowground)	2.8	(34)	1.2	(13)
Company personnel (manual, aboveground)	1.4	(17)	0.8	(9)
Coal operator	0.7	(9)	0.0	(0)
Total	100.1	(1,234)	99.9	(1,073)

Source: Iowa, Census of Iowa, 1905 and 1915, Manuscript Population Schedules for Bluff Creek Township, Monroe County.

Note: Retired miners and "counselors in mines" have been excluded, $N = 9$. Percentages may not sum to 100 because of rounding.

The relative proportions of miners and company personnel shown in Table 3.1 are considerably different from those experienced in many other Iowa coal mines of the period. In smaller mines throughout the state, miners usually accounted for only 70 to 80 percent of the total mining work force. There are a number of reasons that might account for the higher proportion of miners in Buxton. First, Consolidation's mines were the most modern in the state and therefore were more mechanized than most other Iowa mines.[8] For example, because Consolidation's mines were electrified, company personnel transported coal cars by an electrified motor car along the main haulageway to the cage. While it took one company man to operate the motor car that vehicle could pull a far greater number of coal cars than a single mule. As a result, the number of mule drivers should have been less in Buxton than in other Iowa mines, thus reducing the total number of company men. A second reason for the low proportion of company personnel may derive from the state census that was the source of the data. When asked what they did for a living, it is conceivable that many Buxton workers simply stated "miner," as opposed to stating specifically that they were a mule driver, trackman, or cager. Even if the worker stated a specialty, the census taker may have simply written down "miner." It is also probable that the listing for some occupational groups is incomplete, because not all of Consolidation's employees necessarily lived in Bluff Creek Township, the only township from which data were collected for this study. This township included all of Buxton proper and most of the main suburbs. According to a number of former residents, however, some Consolidation employees also lived on acreages and lots located in neighboring townships. Because of this situation, census listings in some occupational categories could be deficient.[9]

Underground company employees—classified as "company personnel (manual, belowground)"—included timbermen, couplers, cagers, tracklayers, trappers, mule drivers, motormen, electricians, and shot firers. Each of these company men had a particular duty. Timbermen, for example, timbered the miners' main entryway to shore up the roof, while couplers coupled empty cars. Cagers pulled empty cars off the cages and pushed on loaded ones. Tracklayers put down railroad track in the main entries and into the individual rooms, while electricians wired the mines for electric power. Trappers were the youngest employees, often only young boys. Their responsibility was to open and close the doors that directed the flow of air through the mine. Each time a mule driver approached, a trapper opened the door so the mule driver could go through without interruption.

Mule drivers drove the mules that pulled loaded coal cars from individual rooms to the main entries, where the motormen came by with electrified motor cars to collect them and transport them to the cage. Mule drivers also had the responsibility of picking up empty cars and delivering them to the individual miners. The rule existed in all Iowa mines that the mule drivers had to "keep an even turn." That meant that they had to deliver the same number of empty cars to each coal miner. This precluded any favoritism toward certain miners on the part of either the company or the mule drivers. A mule driver's work was complicated by the fact that Iowa coal seams were rarely horizontal. One Iowa mining official wrote: "in the first place, the coal beds are irregular in thickness and slope. The beds [seams] roll and pitch to such a degree that the average haulageways in the mines resemble in plan and profile a roller coaster speedway." As a result, mule drivers were constantly either hauling uphill or downhill, with the latter requiring great skill and concentration. In going down a steep incline, drivers had several ways they could slow down their load. They could insert sprags, or sticks, in the wheels of the cars, or they could sand the tracks to give more friction. If they could not slow

Mule driver in a typical pose, with one foot on the chain connecting the mule to the coal car and one hand on the mule's rump. Note the carbide lamp on the front of the miner's cap. (*Courtesy of Iowa State Historical Department*)

down the load, their only choice was to ride it out, hoping that the car would not jump the track.[10]

In a large operation like Consolidation's, the company also hired men to take care of the mules. These men had the responsibility of training the mules when they were first purchased and feeding and caring for them. Also because of its large size, Consolidation kept all mules stabled permanently underground. This practice required a number of men to feed the animals and clean the stables on a daily basis. The stables were usually located near the cage. Archie Harris remembered that at Consolidation Mine No. 18, the company also employed a blacksmith full-time to shoe the mules.[11]

Among company employees, most miners agreed that the shot firer had the most dangerous occupation. The shot firer's responsibility was to walk through the mine twice each day. On the first trip, as mentioned earlier, he visited each room and inspected the holes drilled by the miners. On the second trip, made after all employees had left the mine, he lit the fuses that set off the powder blasts. Each miner was to leave a sufficient length of fuse so that the shot firer could be some distance away before the shot went off. If the miner drilled the holes properly and tamped the powder correctly, as the shot went off it directed the force of the blast back into the coal seam. This resulted in the loosening of large chunks of coal from the face. If the miner had not placed the shot properly, the force was directed into the room. This was called a "misplaced shot," or a "windy shot." The windy shot might cause only a strong gust of air to pass through the mine, or it could cause a current of air powerful enough to kill the shot firer.[12]

According to interviews with former Buxton miners, most men wanted to be miners rather than company men, because they believed that they could make more money by digging coal. Company men worked for a set daily wage, while miners were paid for each ton of coal that they loaded out. Therefore, if miners had helpers (increasing the number of coal cars they received daily) and worked at a rapid pace, they could earn more money than the highest-paid company men (other than those in management). In 1910 a visiting newspaper editor quoted a black Buxton miner who had worked for Consolidation for twenty years as saying: "I have made at least $1,000 every year I have been with them. Many other miners have done as well." Census data at least partially supports this anonymous miner's claim. According to the 1915 state census, thirteen other Buxton miners made $1,000 or more in 1914: seven miners earned $1,000, three earned $1,100, two earned $1,200, and one miner earned $1,500. At the same time, eight miners earned $900 in 1914. Out of this total of twenty-one miners, five were black.[13]

The *Bystander* and the oral interviews conducted for this study also provide information on wages. In 1907, for example, the *Bystander* noted that some miners drew "after their expenses are taken out, as high as $25, $50 and $75 for each two weeks' pay." Also, Robert Wheels recalled that he wanted to dig coal rather than work as a company man because he could make more money as a miner. He remembered that around 1920 some company men at No. 18 made $5.20 per day but that he could almost double that amount as a miner if he could get a continuous supply of coal cars. Hucey Hart recalled: "Old man Jim Washington and Pete Moore, them two every two weeks, it [wasn't] nothing for them to draw $80, $90, or $100 and sometimes even more every two weeks. They'd make more money than the company men would. The company [men] was only getting $7.50 . . . a day. Trapper got $4.00, all the company men got $7.50."

When Jacob Brown and his father worked together in a room, they averaged $60 every payday. Brown recalled that some men made as high as $100 every two weeks. Herman Brooks, who worked as a miner with his father in his midteens, recalled that his father would draw $45 or $50 every two weeks. "That was big money when you could do that." Brooks then drew between $35 and $40 every payday. He explained that even though he and his father did an equal amount of work, he wanted his father to draw the bigger paycheck because his father had a family to support.[14]

The exact wage earned by coal miners and company men in Buxton is difficult to determine. It is possible that informants had a tendency to remember the largest amounts they earned and to forget the many weeks when they earned little or nothing. The state census of 1915 was the only one in which individuals were asked about their earnings. The federal census of 1910 did not include that data. In 1915 census takers asked each individual what his or her wages had been for the previous year. Table 3.2 shows the average income of selected workers in Buxton in 1914. Coal miners in Buxton averaged $499 in 1914. That figure should be viewed with some caution, however. In 1914 the entire Iowa coal industry suffered a reduction in coal orders that affected both railroad mines and independent mines. The *Bystander* noted that Buxton miners experienced long layoffs in that year. In fact, some miners remained unemployed for almost six months. In Buxton, according to the state census of 1915, only 27 percent of white miners and 18 percent of black miners reported full-time employment in 1914. For men who listed unemployment, the average time spent out of work was four months. It is probable that Buxton miners earned considerably less money in 1914 than they had earned in previous years. On the other hand, it is probable that the wages of company men, both management and manual workers, were less affected by the coal slump than were the miners' wages, because company men were paid full wages each day regardless of the number of cars the miners loaded out.[15]

Data on the wages paid coal miners in three incorporated Iowa coal communities during 1914 are also available. In that year, miners in Seymour, Cincinnati, and Beacon—all in southeastern Iowa—averaged $456. This means that Buxton miners earned $43 (9 percent) more than miners in these camps during the year. One explanation for the Buxton miners' higher income is that, as discussed earlier, the Chicago and North Western Railroad provided a steady, year-round

Table 3.2. Annual Income of Selected Workers, Buxton, 1914

Occupational Group	Worker and Wage
Professional and semiprofessional	Physician, $3000; lawyer, $1600; dentist, $1200; secretary of YMCA, $1100; minister, $980; teacher, $490
Business	Merchant, $1060; restauranteur, $1000; manager of telephone company, $900; store clerk, $700
Artisan	Blacksmith, $925; carpenter, $712; shoemaker, $632
Coal production	Top boss, $1500; mine engineer, $960; mule driver, $900; miner, $499
Transportation and communication	Railroad worker, $1162; mail carrier, $1000; telegraph operator, $700; section hand, $400
Service work	Cook, $675; barber, $355; midwife, $250
Laborer	Machinist, $795; teamster, $603; day laborer, $462; laundress, $375; domestic, $201
Agriculture	Farmer, $900; farmhand, $260

Source: Iowa, Census of Iowa, 1915, Manuscript Population Schedules for Bluff Creek Township, Monroe County.

Note: Where categories included more than one worker, the reported income figures were averaged.

demand for coal from the Consolidation mines. Even in the summer, miners could expect at least three days of work each week. Most Iowa coal mines, in contrast, supplied coal only for heating local homes and businesses, and thus they usually shut down each year in April and did not reopen until October. During the summer the miners were forced to look for other employment as, for example, railroad section hands or farm laborers.[16]

An additional consideration in the area of income is the presence of multiple wage earners within households. If a household in Buxton had only one wage earner, in most cases this was the male head of household employed by Consolidation. But if a household had two or three wage earners, it usually included one or two working children. Frequently the children, usually teenagers or young people in their twenties, were also employed by Consolidation. In Buxton in 1914, 24 percent of the black households and 28 percent of the white households reported having more than one wage earner, though the money women earned from boarding, sewing, midwifery, or the sale of dairy or garden produce was not included in the state census of 1915. While the types of jobs held by women and children are discussed in detail in Chapter 5, it should be noted here that the income from wives and children frequently doubled and sometimes even tripled the household's income. In this sense, when considering family income it seems more accurate to consider total family income, rather than income from heads of households.[17]

In some cases, both interview and census data can be utilized to gain a fuller understanding of family earning patterns in Buxton. Alex and Agnes Erickson, second-generation Swedish-Americans, were interviewed for this study. According to the interviews, both went to work for Consolidation while still in their teens and while both lived with their parents, Charles and Josephine Carlson. Agnes, the oldest daughter, worked in the company's business office preparing employees' wage statements, while Alex worked as a coal miner. For most of his time in Buxton, Alex worked in the same room as his father because his father's health was not good. According to the state census of 1915, the Erickson family had a combined income of $1,760 for the previous year. Agnes earned $600, while the father earned $500 and Alex earned $300. Dena, the second daughter, is listed as a clerk earning $360. Through the interview data, it is known that Dena also worked for the company, although obviously in a less important position than her sister. The interview data also reveal that the Erickson family lived in East Swede Town, where they owned their own home (having moved it from Muchakinock in 1900) and paid the company fifty cents per month for lot rental. Although both Alex and his father had fees deducted from their wages for union dues and other expenses, it seems that the Ericksons' living expenses were modest. In other words, the family had a considerable amount of money to spend on nonessential items.[18]

The 1915 census also allows for a detailed view of black family income in Buxton. In 1915, Mary and Wiley Parker lived in Buxton with their seven children. The six oldest offspring carried the last name of Mayes, indicating that these children were from an earlier marriage of Mary Parker's. Mary and Wiley apparently had one child of their own, as the youngest child in the household was eleven-year-old Olivia Parker. The Parker household contained seven wage earners. Although Wiley Parker listed no occupation, he earned $400 in 1914. Four sons were coal miners earning a combined income of $2,200. Another son earned $700 working as a waiter, while a daughter earned $200 as a domestic. Altogether, family members earned $3,500. Two sons who worked as miners and who earned a combined total of $1,000 were under twenty-one, and it is probable that their father collected their total income. The incomes of the older children were no doubt shared with the parents and contributed to the family's general living expenses. While the Parker family earned more than most black families in Buxton in 1914, there were many other cases where black families earned between $2,000 and $3,000. In many of these cases, all working family members were employed by Consolidation. Given the high standard of living and the material comforts that must have flowed from these wages, it is not difficult to understand why blacks held Consolidation in such high esteem. Particularly for families who had several offspring employed by Consolidation, the company could do no wrong.[19]

Finally, a comparison can also be made between the earnings of company employees who did manual work and the earnings of Buxton residents in the professional or managerial class. The 1915 state census lists Dr. E. A. Carter's 1914 income as $3,000. Another company employee, Eric Brown, a cashier for Consolidation in charge of payroll, earned $2,500. George Woodson, a black lawyer in Buxton, listed his 1914 income as $2,000. Undoubtedly the incomes of these three people were viewed as sizeable in 1914. In these households, only the head of household (Woodson was single) was employed. It is instructive to note that the families of well over two dozen coal miners and company men (who were classified as manual workers) earned as much as or more than Carter, Brown, or Woodson.[20]

District 13 of the United Mine Workers of America and the Iowa Coal Operators' Association determined the wages of all mine employees other than management personnel. Beginning in 1898, the UMW and the Iowa coal operators entered into either a yearly or biennial agreement known as the "Joint Wage Agreement," which determined the wage to be paid in Iowa's coal mines during the period. In 1910, for example, the Joint Wage Agreement stated that miners in Sub-district 2, which included Monroe County, would receive $1.00 per ton for coal loaded. Tracklayers, timbermen, cagers, and mule drivers received $2.70 per day.[21]

The wages drawn by the men on payday, however, did not represent their full earnings. Before issuing wages, company bookkeepers subtracted house rent, lot rent (if applicable), company store charges, water and coal delivery fees, and fees accrued by the men during their work. These included blacksmithing fees for sharpening tools (each miner paid one percent of his gross earnings), a fee of $1.00 per month for train transportation, union dues, and $1.00 or $1.50 for medical dues. A major expense incurred in the course of the miners' work was the purchase of blasting powder, which they charged at the company store. Consolidation, like most coal firms in Iowa, deducted union dues and medical dues at the request of the miners.[22]

Although coal miners could make more money than company men, they did not have the same guarantee of wages. While coal miners could do well under normal conditions, they might encounter "dirty coal," which contained sulphur, dirt, or rock. Companies did not allow miners to load "dirty coal." If a miner encountered an inordinate amount of rock or dirt in his room, he could talk to his foreman and try to work out an arrangement whereby he would be paid something for digging out the rock. There was no guarantee, however, that the foreman would grant the concession. Also, once a miner's room had been mined out, his name went on the foreman's list for assignment to a new room, and there was frequently a wait before a new room was opened, during which time the miner earned no money. The waiting period could extend over a period of several months.[23]

Throughout its existence, Buxton was the home of mine employees from many different ethnic backgrounds and with widely differing degrees of mining experience. Many of Buxton's Swedish miners had moved from Muchy, where they had also worked as coal miners. Moreover, Buxton contained miners of English, Welsh, and Scottish descent. Many of these men had previously worked as miners in other states and had come to Iowa with considerable mining experience. In fact, some of these men had learned the mining trade in their native lands. But it is also true that Consolidation continued to recruit black workers from Virginia through at least the first decade of the twentieth century, and interviews indicate that few, if any, Virginia blacks had mined coal before settling in Buxton. These differences, however, do not seem to have caused friction or produced conflict between members of different ethnic groups or between older and younger members. Nor did interviewees remember any segregation of work on the basis of ethnicity or race, or any discrimination by company officials. Most men interviewed had been miners and, because of the independent nature of coal mining, with men working individually or in pairs in isolated settings, the men's main concerns centered on their own rooms and the total amount of coal they were able to load. In other words, concerns centered around individual performance. Interpersonal relations seemed to be less important to coal miners than to many other types of workers.[24]

The occupation of coal mining in Iowa, as elsewhere, was considered highly dangerous. The major cause of death and injury was falling slate. The careful miner always propped up loose ceiling slate, but sometimes slate fell without warning from an area that did not appear to be loose, or a miner might accidently knock out a prop and cause an immediate fall of slate. Some miners suffered blindness as a result of sulphur or coal striking them in the eye. After the introduction of mining machines in Iowa mines, machine operators had to be constantly alert not to get their clothing caught in the machine and have an arm or leg pulled into the rotating blades.[25]

While Buxton miners faced certain unalterable conditions in their work underground, they also soon learned that some conditions could be changed. From 1900 to 1923, Buxton miners were the recipients, as were all Iowa miners, of many positive changes brought about by District 13 of the United Mine Workers of America. During the first quarter of the twentieth century, officials of District 13 worked to secure a series of occupational, medical, educational, and social benefits for Iowa's coal mining population. District 13 was first organized in 1891, but in the mid-1890s, due to the effects of the depression of 1893, the district organization disbanded. In 1898, Iowa miners reorganized the district and from then until the early 1940s District 13 remained a strong force in the lives of Iowa's miners, promoting reforms and programs that greatly improved the quality of life for both coal miners and their families. Buxton miners joined the union shortly after the community was established, creating Buxton Local 1799. The Local remained formally organized until 1927, and it is probable that miners at Haydock continued their affiliation with the UMW through Buxton Local 1799.[26]

During the initial years of District 13's existence, officials thought mainly in terms of wages and hours. They achieved considerable success in these areas, as by 1900 they had secured the eight-hour workday and had established the principle of the joint wage agreement. The latter in effect stabilized wages and provided the union with an orderly method of requesting wage increases. Following these two victories, District officials continued to press for other economic changes that they believed would eliminate restrictive company policies. In the early 1900s, District officials won the right to have a shot examiner and shot firer in every mine (usually the position was combined). This meant, first, that blasting was done by a qualified person and, second, that the blasting was not done until all miners had left the mine, thus eliminating the danger of a windy shot while the miners were at work. Early in the twentieth century, District 13 officials also won the right to place a union checkweighman at every mine. For many years miners had insisted that company weighmen did not always give them credit for their total tonnage. With a union checkweighman working alongside the company weighman, the miners believed that their interests were safeguarded. The elimination of coal screens marked an additional concession won by District 13 officials. For years, coal operators had insisted on putting all coal through a series of screens that eliminated the smaller chunks. Only the larger pieces left on top of the screens were then weighed, and coal operators did not pay miners for loading out the coal that passed through the screens. Not until 1918, however, did all Iowa coal operators finally agree to eliminate the screens.[27]

Beginning in 1905, District 13 officials began to turn their attention to those areas not affected by bread-and-butter issues. Their first major concern was the matter of death benefits. In 1906 delegates to the union's annual district conference voted to give $100 in death benefits to the widows or other legal heirs of deceased members. A few years later, the district membership raised the death benefit to $200. Frequently the fellow workers of an injured or deceased member would also take up a collection for the deceased's family, and this could amount to several hundred dollars. When Odessa Booker's father was killed in a mining accident the other miners collected around $500, which they gave to Odessa's mother.[28]

Within five years, District 13 also had a hospitalization plan for miners and their families. In 1911, Dr. R. P. Miller of Albia proposed making hospital services available to coal miners and their families. In that year, Dr. T. E. Gutch offered each miner a contract for twenty-five cents per month whereby the miner had coverage for hospitalization, including surgery. District 13 president John White had earlier suggested such a plan, and following the establishment of Gutch's hospital, the District 13 leadership heartily endorsed the project and urged Iowa miners to join. Within a short time, the plan was extended, at a cost of fifty cents per month, to cover the miners' families. One Albia

Company physicians in Buxton sitting in front of their office. Seated on the chairs, from left to right, are Dr. Edward A. Carter, Dr. Powell, and Dr. Gray. The young man on the far left is Jim Warren, brother-in-law of Dr. Carter. (*Courtesy of Iowa State Historical Department*)

resident described the facility, known as the Albia Miners' Hospital, as an "up-to-date hospital with the latest modern equipment in every department, representing an investment of over $20,000, so that the miner knows he has everything at the best of his service." Five years later, Dr. Gutch established a second hospital facility for miners and their families in Des Moines. The latter facility followed the same plan as the Albia hospital and charged the same fee of fifty cents per month. Nellie King remembered that her husband, Charles, went to the Albia Miners' Hospital for an operation on his liver. The cost of the operation was covered by their monthly hospitalization fee.[29]

The hospitalization plan supplemented a medical plan that had been in effect for many years in Iowa's camps. Although it had no official connection with the UMW, the practice had long existed in Iowa's coal camps that companies either hired physicians to practice in the communities or made arrangements with neighboring physicians to provide medical care for company employees. By 1900 most coal miners paid a fee of one dollar per month, which provided medical care for themselves and their families. In some camps, the company doctor charged a small fee for delivering a baby (around five dollars), but other medical treatment, including prescriptions, was included in the monthly fee. In Buxton the company hired either three or four physicians. In 1907, for example, Consolidation employed Dr. J. H. Henderson as chief physician, Dr. James Muir as assistant chief physician, and Dr. E. A. Carter as assistant physician. Drs. Henderson and Muir were both white. Although Consolidation deducted the medical fee from their employees' wages, these fees were passed along directly to the physicians.[30]

In general, former residents had high praise for company doctors, particularly Dr. E. A. Carter. Moreover, they believed that the medical care they received in Buxton was of high quality. At times, however, a lack of medical facilities and medical knowledge was evident. Herman Brooks recalled that difficulty in getting proper treatment in Buxton cost him the possibility of starting a new career. Brooks, a coal miner and a member of the Buxton Wonders baseball team, related that around 1910 he received an offer to pitch for an all-black Chicago baseball team, the Chicago White Stockings. The White Stockings proposed to pay Brooks forty dollars per week, far more than he earned as a miner. A few days before he was to quit his mining job and leave for Chicago, a chunk of slate fell on his arm, breaking his wrist. He went to see a company doctor immediately but was told to come back in the morning so the doctor could check his arm. Brooks spent the night in great pain and returned to the doctor the next morning. The arm was set but several weeks later Brooks returned to the doctors' office, convinced that the wrist was not healing properly. While in the waiting room on his second visit, Brooks observed a miner seated on a chair in the corner. The miner had been hurt, and to relieve his pain, a doctor's assistant kept passing a bottle of ether under the miner's nose. After a time, the assistant discovered that the miner was dead. Brooks believed that the miner had died of an overdose of ether. A short time later, after several doctors had examined Brooks's wrist, they concluded that it would have to be rebroken and reset. When the doctors suggested giving Brooks ether, he protested. The doctor replied, "Herman, what we gonna do, we got to break it over." Brooks suggested that they go over to the company store and get several miners to hold him down while they rebroke the wrist. Brooks related the procedure:

It was those men on me that held me down. I couldn't move. They had a man on each leg. Man sitting on each leg. They were sitting on the legs. They were sitting across here see and they had me stretched out. They had a table there. They had me stretched out and so he just laid my arm out like that and come down with that rubber mallet and broke it.[31]

Brooks's wrist never healed properly, thus ending the possibility of his career as a professional baseball player.[32] With a few exceptions, however, such as Herman Brooks, company employees and their families in Buxton had good health care, probably better than most Iowans in the early twentieth century.

After 1911, District 13 of the UMW continued to promote a more comfortable and enjoyable life for Buxton's coal miners. An investigation conducted in 1919 by UMW officials, representatives of the Iowa Coal Operators Association, and personnel from the Department of Public Instruction revealed that many camp schools were overcrowded and contained insufficient supplies and textbooks. They cited the Buxton schools, along with many others, as overcrowded and badly in need of additional teachers, more classrooms, and better textbooks. As a result, from 1919 through the 1920s, the state legislature appropriated money for the upgrading of the camp schools. At approximately the same time, Iowa UMW leaders began to push for an adult night-school program for foreign-born miners and their children. In 1915 the state census listed 273,484 foreign-born persons in Iowa, 37,169 of whom were unable to speak English. District 13 officials had a particular interest in establishing adult educational programs because a large percentage of the foreign-born lived in coal mining communities. District officials believed that not only would many of the foreign-born take advantage of school programs, but that an additional 4,000 foreign-born miners' sons would also respond. When a bill for the night-school project reached the General Assembly, however, it failed to pass.[33]

Because the records of Buxton's Local 1799 have not survived, it is not known who served as the local officers or the degree to which black and white members participated in the union. It is, however, evident from other sources—such as the *United Mine Workers Journal* and the District 13 Executive Board Proceedings—that at least some blacks served as officers in the Buxton local, and that they served as delegates to district and national conventions. In January 1912 the *Bystander* reported that at the district convention G. D. Yancey, E. P. Thomas, and W. K. Brown, all black miners, had been chosen as delegates to the national convention of the UMW, which would be held in Indianapolis. In at least one case, a black miner, G. D. Yancey, was elected to the scale committee (a standing committee) of District 13.[34]

Black miners in Buxton sometimes used the UMW as a forum through which they could speak out against racial discrimination and as a vehicle by which they could advance black interests. In 1906 a Buxton delegate offered the following resolution at the District 13 convention in Des Moines:

> That we will not discriminate against a brother on account of creed, color, or nationality, and that we will defend freedom of thought, etc., and, in view of the fact that about one-third of the members of the U.M.W. of A. are colored and do pay about one-third of the taxes and revenue. Therefore, Be It Resolved, That the colored mine workers have more representation on the official staff, as we know that taxation without representation will not keep any people satisfied long.[35]

It does not appear that the Convention took any action on the resolution. Four years later Buxton delegates again submitted a resolution to the District 13 annual convention dealing with racial discrimination: "When an employer refuses to employ a brother on account of his color or nationality, other brothers shall refrain from seeking employment at that place." The resolutions committee concurred with the resolution and recommended that it be "placed in the hands of the Scale Committee at the Convention . . . in order to have it entered in the next [joint wage] agreement." Unfortunately, the records do not include any details of this action, so it is unknown what particular employer the blacks had in mind. It is certain, however, that the resolution was not directed against Consolidation. As with the earlier resolution, it does not appear that the delegates acted on the matter.[36]

In 1911 black miners in Buxton made another effort to speak out against racial discrimination. That action involved 160 black miners, all union members in good standing, who left Buxton and Oralabor, a coal camp in Polk County, and applied for jobs at Ogden. Earlier that year, the white Ogden miners, working in an all-white coalfield, had gone on strike against a company coal screening policy. Union officials quickly informed them that they were subject to fines if they carried out an unauthorized strike, so the miners returned to work and resigned. A short time later, the 160 black miners were hired to work in the Ogden mine. Immediately after the blacks arrived, the white miners changed their story, insisting that they were only on strike, thus casting the blacks in the role of strikebreakers. In the meantime, the company had evicted the white miners and turned the company houses over to the black miners and their families. The situation was settled a year later when the black miners returned to Buxton and the white miners reclaimed their old jobs.[37]

While there is no testimony available from black participants regarding their motivation for moving to Ogden, it seems to have been at least in part racially motivated. While nothing is known about the background of the Oralabor miners, it seems that the Buxton miners and their families were leaving certain employment, comfortable housing, and a lack of racial discrimination. In moving to Ogden, the Buxton blacks were apparently opting for temporary employment and the possibility of a hostile reception on the part of white miners in an all-white community. They did, in fact, encounter hostility. On one occasion a black man was badly beaten by several white miners, and later a black man was accused of raping a white woman, although the latter case never reached a court of law.[38] While it cannot be completely discounted that the black miners went to Ogden only in search of employment, it seems more probable that these men saw an opportunity to integrate what had previously been a all-white coalfield. While they did succeed in the short run, they eventually returned to Buxton and resumed their positions there.

While the general benefits secured by the UMW certainly made the lives of Buxton's miners more comfortable, the coal industry itself possessed certain occupational characteristics that provided miners with additional security. In the area of housing, employees were virtually guaranteed a company house regardless of whether they were able to pay the rent on a monthly basis. If the company experienced a slowdown and miners worked only two days a week, or perhaps two weeks a month, the employees continued to live in the company homes. In the event that a miner was waiting for a new room to be opened, he and his family remained in the company home with the understanding that when he did go back to work the past rent would be deducted from his wages. During this time the miner was also free to continue charging at the company store.[39]

Perhaps the most unusual aspect of employment for coal miners was the fact that coal companies rarely laid off employees. In the event that a company wanted to cut production because of a reduction in orders, rather than firing a number of employees the company simply cut back the work schedule to only one or two days per week. This policy obviously helped the company in that these miners continued to live in company housing and continued to trade at the company store. It also helped the miners because they all retained their jobs, although they made less money. Coal companies, moreover, rarely fired employees for a violation of rules or poor performance. Company officials supervised certain areas of the mines closely, such as the entryways and haulageways, where the company took responsibility for upkeep, but what a miner did in his own room was viewed as the miner's business. If he worked hard during the day and loaded out all his assigned coal cars, that was considered typical. But in the event that a miner became tired and decided to stop work before loading his final car, that was the miner's concern, and he simply made less money that day. Alex Erickson recalled that Consolidation did not fire people often but "if you got fired or something, there was always little mines around there. You could go up there and get a job and work there for a while, then you could come back and ask for a job again and they'd hire you." Erickson added that the company might fire a miner for loading dirty coal (coal with dirt, sulphur, or other impurities) or if "a fella was careless with his mule, and his mule got killed. . . . They thought more of the mule than they did of the driver, so the driver he'd go to another place and drive for a while and pretty soon he'd drift back."[40]

Buxton miners, like miners everywhere, experienced a distinct form of discrimination brought on by their work in the coal industry. Throughout the nineteenth century, coal miners in all parts of the country experienced the stigma of social inferiority. Many Americans held the belief that coal miners were improvident and intemperate. A common nineteenth-century characterization of coal miners was that they "live hard, drink hard and die hard." Moreover, miners by necessity moved frequently, and this cast doubt on their ability to be good citizens. Coal operators had long insisted that miners were "good spenders" and that they did so purposely to maintain their popularity.[41] This social stigma carried over into Iowa. Many men and women who grew up in coal camps recalled vividly that they were regarded as second-class citizens by non-coal-mining families. Further intensifying this occupational discrimination was the fact that after 1900 Iowa's coal camps contained large numbers of southern and eastern Europeans. The old view persisted that somehow these people were inferior to the earlier immigrants from northern and western Europe. An Italian-American woman raised in a series of Iowa coal camps remembered that many families outside the camps were farmers and Protestants: "We were coal miners and Catholics and Italians to boot—we had three strikes against us."[42]

Highbridge, Dallas County, was typical of Iowa coal camps in the 1920s. Notice how close together the houses are. The company store is the large building on the left. (*Courtesy of Iowa Mines and Minerals Department, Des Moines*)

The view also persisted that, while coal miners were less desirable citizens than non-coal miners, the coal miners who had first immigrated were somewhat more acceptable. Coal miners from England, Wales, Scotland, and Sweden were held to be in a higher social class than the later-arriving southern and eastern Europeans. The former had frequently learned the mining trade in their native countries and were regarded as craftsmen, not simply as miners. The arrival of the eastern and southern Europeans, on the other hand, coincided with the changeover from pick or hand mining to machine mining. This led many coal operators, both in Iowa and in the eastern states, to dismiss the latecomers as men with strong backs but no skills. Implicit in the operators' thinking was that the southern and eastern Europeans were not skilled craftsmen but simply common laborers.[43]

Coal miners and their families in Buxton experienced some degree of discrimination because of their occupation, but overall this was mitigated because of the town's unusual characteristics. Buxton's large size provided a degree of insulation and, in turn, independence from outside populations. Local residents were not required to travel to neighboring communities to shop and to secure medical services. Nor did Buxton residents find it necessary to travel to other communities for social activities. In fact, the opposite was often true: residents of surrounding communities frequently traveled to Buxton for shopping and to attend events at the YMCAs.[44] These factors did not erase the particular mentality concerning coal miners held by many rural and small-town residents, but it did mean that Buxton residents were exposed to this mentality less frequently.

Buxton's large black population also provided its members with a sense of security that tended to mitigate discriminatory attitudes. Lola Reeves related that after living in Anamosa, where her family was one of three black families in town, she experienced a very different situation in Buxton. Mrs. Reeves had been the youngest child within the three black families, so during her last five years in school she was the only black child in town. She stated:

> Consequently, I had been raised in a white surrounding. Going to Buxton with all the people of my own race was a great experience for me. I learned a lot and I acted shy and timid [at] first. But after I got there, I could exercise my feelings, my potentials, my talent and my social life and I think Buxton brought a whole lot of joy to me, just to be able to live and, a colored girl, in a colored area and feeling like I was one of them and I was happy.[45]

The view certainly persisted in the minds of the non-mining population living in and around Buxton, however, that miners there, particularly blacks, lived well (often beyond their means), dressed well, spent large sums on drinking and gambling, and consequently saved little if any money. In this sense, the old attitudes toward coal miners as improvident and intemperate people continued to live on in Buxton.

. . .

For Buxton mine employees, black and white alike, the Consolidation Coal Company proved to be a good employer. While there were certain aspects of mining that the company could not eliminate, such as its dangerous nature, there were many conditions both above- and belowground that the company could make more tolerable. Since the Buxton mines were the most modern in the state, presumably they were also among the safest. Working in a modern mine should have been a more pleasant experience than working in one without electricity. At the same time, the coal company also controlled life aboveground. Through its construction of good homes, excellent shopping facilities, and community facilities, Consolidation produced a highly satisfied work force. Consolidation's employees, moreover, enjoyed all the advantages offered by the UMW, and by 1920 these advantages were considerable in both the economic and social areas.

Looking more closely at Consolidation's labor force, there were other factors which also seem positive. In many families, not only did the male head of household work for Consolidation, but often several other family members did so as well. In some cases these different jobs produced a sizeable family income. Perhaps of equal importance is that jobs were available for most family members who wished to work. On the average, Consolidation's employees experienced less unemployment—and therefore more income—than most other Iowa mine employees. Good wages, steady employment, and pleasant, convenient surroundings went a long way toward producing satisfied workers in the company town of Buxton.

4

THE CONSOLIDATION COAL COMPANY

OR residents of Buxton, life began and ended with the Consolidation Coal Company. Because Consolidation founded the community and because Buxton remained a company town throughout its existence, Consolidation made the decisions that either directly or indirectly controlled all aspects of community life. In light of this pervasive authority, it is necessary to examine the company not only as an employer, but also as a community manager.

Because of the period in which Buxton existed, it is also necessary to examine company policies and practices from another perspective, that of welfare capitalism. During the time that Consolidation operated in Iowa, from 1881 to 1925, a movement known as welfare capitalism existed in many parts of the country. This movement was one where leading industrial firms provided their workers with services and facilities designed to produce a more satisfied work force, a move that industrialists hoped would stave off industrial unrest and unionization. Because the period of welfare capitalism coincided with Consolidation's activities in Iowa and because of the extensive nature of the movement, Consolidation's policies will be examined from that perspective. One aspect of Consolidation's experience is clear at the outset, however: former residents held the company in high regard, and as employees they followed company policies without question.

By the fall of 1900, when construction began on the first homes in Buxton, Consolidation had done considerable planning as to the type of community they intended to create. In 1896, when Consolidation officials were beginning to look for new coalfields, numerous industrialists in the eastern United States were following a set of practices known as welfare capitalism, which Stuart D. Brandes has defined as "any service provided for the comfort or improvement of employees which was neither a necessity of the industry nor required by law." These services took a variety of forms, including the construction of housing, parks, libraries, schools, churches, YMCAs, and medical facilities. Industrialists provided these services or facilities to produce a more contented work force; more specifically, the industrialists hoped to prevent union organizing and labor unrest, particularly strikes. Accordingly, Brandes notes, "the underlying theory [of welfare capitalism] was that well-housed, well-fed, clean, properly educated Christians do not strike, or at least were less likely to than those with different lifestyles."[1]

Welfare capitalism first appeared in the railroad industry. As early as 1872, eastern railroads began to build YMCAs as a way of avoiding labor disruptions. The YMCAs contained bathing facilities, libraries, and athletic facilities; moreover, "Y" officials offered classes on railroad work and religion. In 1877 a series of strikes paralyzed many eastern railroads, and some railroad executives responded by upgrading workers' living conditions, believing that these improvements would prevent workers from striking again.[2] As the philosophy of welfare capitalism spread, some industrialists constructed entire towns based on its principles. In 1880, for example, George Pullman constructed his model community of Pullman, Illinois, on three hundred acres near Chicago. The town housed Pullman's labor force, which manufactured Pullman sleeping cars. In addition to fine houses, many of which were five-room row houses, Pullman built parks and a miniature lake for boating and swimming, complete with an island to be used for different types of athletics. Workmen paved Pullman's streets with macadam, built wooden sidewalks, and landscaped the entire area. By 1885, Pullman contained 1,400 dwelling units. Pullman's project is viewed as the first model town in industrial America.[3]

By improving the lives of workers and their families, welfare capitalists believed they could accomplish several goals. Beyond preventing their workers from joining unions and engaging in any type of industrial violence, welfare capitalists believed that through their upgrading of workers' lives, they could actually create an "improved American working man: thrifty, clean, temperate, intelligent, and, especially, industrious and loyal." By accomplishing this goal, they would not only fight the expansion of unionism, but perhaps create an environment where the idea of unionism would not breed at all. In their development of welfare capitalism, industrialists placed heavy emphasis on providing good homes. They believed that comfortable homes surrounded with trees and flowers kept families intact and in turn produced satisfied, productive workers. Moreover, some companies sponsored bands and baseball teams. They reasoned that if these teams were successful they might create pride and loyalty in all the workers, not just in team members.[4]

It is not known to what extent either John Emory Buxton or Ben C. Buxton was influenced by Pullman or other welfare capitalists. Certainly there are no visible direct links—such as business records or private correspondence—between the Consolidation Coal Company and other individuals or companies involved in the movement. There is every reason to believe, however, that the Buxtons were aware of welfare capitalism. The Chicago and North Western Railroad was headquartered in Chicago, and the Buxtons and other Consolidation personnel maintained close connections with the main office. Although Ben Buxton was only twenty-five years old in 1896, when he became general superintendent of Consolidation, his father had been associated with the railroad from the early 1880s. It is certainly likely that the elder Buxton, an easterner by birth, would have known of the railroads' practice of building YMCAs and other institutions for their workers.[5]

Regardless of whether Consolidation officials consciously set out in 1896 to imitate the actions of men like George Pullman and to follow the tenets of welfare capitalism, that is in effect the path that Consolidation followed. By building substantial homes, creating parks, erecting extensive modern facilities, and constructing two YMCAs, the company produced a community quite akin to Pullman and other company towns established under the influence of welfare capitalism. In addition, Consolidation officials moved even beyond some welfare capitalists, implementing liberal housing and shopping policies not typically granted in company towns. These policies included giving employees the option of building homes on land rented from the company and trading at other than company-owned stores. This flexibility on the part of Consolidation was also evident in minor ways. For example, Consolidation allowed numerous peddlers to solicit in Buxton, a practice often prohibited in company towns. In fact, some companies erected fences around their towns to keep out unauthorized personnel.[6]

While not all companies that embraced welfare capitalism

managed to eliminate industrial unrest, for the most part the Consolidation Coal Company did achieve that goal. It should be noted, however, that Consolidation officials were not attempting to prevent workers from unionizing. In the fall of 1900, shortly after the first miners moved to Buxton, both the Buxton miners and those remaining in Muchakinock voted to join the United Mine Workers of America. At the time, John Reese, vice-president of District 13, reported to the *United Mine Workers Journal* that he was highly pleased that these two large camps had organized. He did not mention any resistance by Consolidation, and there is no evidence that the company later attempted to break up the union.[7] In general, Consolidation succeeded in producing a highly satisfied work force that rarely, if ever, took exception to company policy or resisted that policy.

It is not clear, however, whether Consolidation produced an "improved American working man" who lived thriftily and temperately. Former residents remembered that, although the company did not allow drinking or the sale of liquor on company property (Buxton proper), businesses that manufactured and sold liquor did exist in outlying areas. Informants also recalled that after employees received their two-weeks' pay at noon on Saturday, some men headed into the nearby timber and gambled steadily until returning to work on Monday morning. Informants also recalled that Des Moines prostitutes came by train to Buxton on the Friday preceding payday.[8]

Buxton did experience some violence among its residents, but this was always of a personal or nonindustrial nature. Informants varied in their memories of violence. In general, people who lived in Buxton remembered that there were only a few murders and robberies and an occasional knifing. On the other hand, people who lived near Buxton but not in the town tended to remember more violence. Incidents usually involved members of the same ethnic group, although there were some interracial confrontations. Regardless of the number of fights and other incidents, however, these did not disrupt the work process, other than an occasional miner missing work on Monday morning. In comparison to other coal mining communities in Iowa, Buxton did not seem to have an excessive amount of violence.

Even if we cannot be certain as to John and Ben Buxton's knowledge of welfare capitalism in 1896, we do know that Consolidation's policies paralleled the practices and policies of eastern industrial firms that were involved in that movement. The most obvious similarity was the physical environment created by Consolidation. If Consolidation

Ben Buxton and his two daughters outside a company building. Buxton served as general superintendent of Consolidation from 1896 to 1909. (*Courtesy of Iowa State Historical Department*)

officials had proceeded in the manner typical of other Iowa coal operators, the new camp would have been laid out in haphazard fashion, almost at random, with little attention given to water availability or drainage conditions. Residents in other Iowa camps continually complained that water collected in pools after every rain, and these became breeding grounds for mosquitoes.[9] Buxton did not have that problem. Planners laid out the town on a gently sloping hill so that almost all parts of the community benefited from the sloping terrain. In fact, the location selected by the company and the physical environment it created indicate that from the beginning Consolidation officials intended Buxton to be a very different company town from those found in other parts of the Hawkeye State.

The company housing in Buxton represented another major improvement over that found in most Iowa coal camps. For two decades, Iowa's coal miners had protested the small, flimsy houses provided by their employers. Operators sometimes expressed the view that mining families did not deserve better homes because they would not take care of them. Consolidation's action in building substantial five- and six-room homes seemed to reject that argument. Not only did the company construct better homes than were typically found in coal camps, but they also deviated from the traditional pattern in that they provided continual upkeep for the dwellings.[10] It is difficult to assign a motive for Consolidation's building better housing other than the influence of welfare capitalism. Given the typical camp housing in Iowa and the fact that miners, although they indicated a desire for better houses, had no recourse but to accept small, poorly constructed houses if that was what the company built, it is clear that Consolidation was under no obligation to build better homes. Because the camp of Muchakinock contained typical camp housing, that experience provided no antecedent for a better grade of housing in Buxton. Moreover, Consolidation was not bound to provide better housing to insure a continual supply of workers. Coal miners in Iowa, like other areas, exhibited a high rate of geographical mobility, going where the jobs existed, not where the physical environment was superior. It is interesting to note that former residents talked about migrating to Buxton because that is where they could find jobs. Not a single informant mentioned that they were initially attracted to Buxton because of better housing or other improved physical conditions.

Buxton's business community and public improvements also showed a sharp contrast with other Iowa coal camps. In looking at the physical environment, it is clear that the company intended to establish at least a semipermanent community, as well as to provide their employees with greater social and cultural opportunities. The large size and extensive stock of both company stores provided residents with shopping opportunities unknown in other Iowa coal camps. In addition, Buxton contained over forty independent business operations that had no connection with the Consolidation Coal Company. Some were large ventures—such as the Buxton Hotel, owned and managed by the Gaines family, and Hobe Armstrong's meat market, which was the main meat supplier for the community—but there was

Buxton women, decorated with lodge banners and sashes, gather in front of the Buxton Hotel for a lodge function. (*Courtesy of Donald Gaines*)

also a wide range of smaller grocery, clothing, and notions stores that greatly increased the residents' shopping options. Buxton's shopping facilities were, in fact, more extensive than those found in most incorporated Iowa communities of equal or slightly larger size. The facilities provided by the two YMCAs also improved the quality of life for all Buxton's residents, and during the community's early years the company established at least two parks, several tennis courts, and two baseball diamonds. Two organizations, the Buxton Cornet Band, an all-black organization, and the Buxton Wonders, a predominantly black baseball team, were partially sponsored and enthusiastically supported by the company.[11]

As well as providing comfortable homes and a pleasant environment for their employees, Consolidation also adopted other policies that made their workers' lives more pleasant and economically rewarding. As discussed earlier, beginning in 1900 the company offered to lease any employee either a town lot or several acres of nearby land so that he could construct his own home. Both white and black workers took advantage of this offer. It is interesting to note that other Iowa coal operators would eventually, in the late teens, give verbal support to the desirability of this practice, but it is not known if any other operators actually implemented a leasing program. Employees who accepted Consolidation's leasing option had the opportunity to save considerable rent money, which many then used to modernize or enlarge their homes.[12]

Consolidation's policy of leasing land for private housing no doubt produced a more contented work force but was not done without some expense to the company. Coal companies traditionally constructed homes for workers on which the companies realized a profit. Although many coal operators emphasized that their major profits came from coal production, operating a coal camp was an additional business venture. It was to any operator's advantage to rent as many company houses as possible. In fact, some companies required a prospective employee to sign a rental agreement before they agreed to hire him. In Buxton, Consolidation was certainly passing up some profit by leasing land for fifty cents an acre when the company could have been collecting seven or eight dollars per house per month. There is no indication that a housing shortage existed in Buxton, a situation that might have prompted the company to initiate the leasing program. Rather, many accounts indicate that some houses stood empty at all times.[13]

Aside from practices related to welfare capitalism, it is also necessary to examine Consolidation's employment record in regard to specific personnel practices and its role as a community manager. Throughout the early years of Buxton, the Consolidation Coal Company played a distinctly paternal role in the lives of Buxton's residents. At Christmas, Consolidation and the company store provided every child in Buxton with a present and candy. Alex Erickson remembered that, starting on the day after Christmas, company store officials allowed each church to bring in its Sunday school students, and each church had an evening assigned to it. The company put out the gifts in a special area and each child could choose one. Jessie Frazier remembered that the company also provided each family with a turkey at Thanksgiving. Buxton typically celebrated the Fourth of July with parades, concerts, speeches, and races, and the company store provided food for the barbecue. In July 1911 the *Bystander* noted that the company had provided pork, beef, and mutton. The same year, the company provided the Buxton Band with a new location for its concerts, selecting a site between the company store and Hobe Armstrong's meat market.[14]

Interviews with former Buxton residents almost unanimously support the view that Consolidation was both a generous and a fair employer. Black former residents remembered that Consolidation exercised no discrimination in hiring. The majority of both blacks and whites were hired as miners, but the company also hired blacks as entry men (company men who mined out the main entries in the mine), mule drivers, timbermen, and electricians. Apparently, no blacks were hired as foremen, but it does not appear that blacks were excluded from other company positions. The great majority of blacks hired by Consolidation worked underground, but the company also hired blacks to clerk in the company store. Although Consolidation did not at any time hire a black manager for the company store, they did hire at least one black to manage a store department. Occasionally Consolidation hired blacks in semiprofessional and business positions. William W. Lee was employed by Consolidation as a hoisting engineer from 1893 until his death in 1908. Born in Charlottesville, Virginia, in 1870, Lee had come with his parents to Muscatine, Iowa, and had later moved to Muchakinock. He served as a hoisting engineer at Buxton No. 10. Lottie Armstrong Baxter, daughter of Hobe Armstrong, served as the cashier in the Buxton Bank (one of three officers) for several years. Consolidation hired one black physician, Dr. E. A. Carter, but otherwise hired white physicians to work for the company.[15]

Another consideration in assessing Consolidation's record as an employer is that of the company's assistance to injured or incapacitated employees. Although the lack of company records precludes any systematic survey of company assistance, interviews with former residents indicate that Consolidation provided aid for at least some injured employees. Sometimes that aid violated state law. Robert Wheels related that he went to work with his father in the mines at age fifteen because his father became ill and could not work alone in his mine room. According to Wheels, the company "put a year on my age" so that he could work with his father. Since state law required young men to be sixteen years of age before working underground, the company had broken the law. The result, however, enabled the father to continue working, thus providing an income for the Wheels family.[16]

The company sometimes aided widows whose husbands were killed in the mines. Susie Robinson related that her husband, a mule driver, was killed in a mining accident. The company, she said, "gave me so much money. I got $4,000, . . . but they didn't give it to me all at once. They just paid me a monthly pay." Mrs. Robinson stated that in addition to receiving money she did not have "to pay for nothing, I didn't have to pay only just to buy my food. They paid for everything." In effect, the company allowed Mrs. Robinson to live in a company house rent-free and provided her with free coal. The company offered Mrs. Robinson a job in the company store but she felt that with three small children to care for she could not manage an outside job.[17]

Odessa Booker recalled that when her father was killed in a mine accident the company responded by finding employment for both her and her younger brother. She was fifteen and her brother was fourteen at the time, and, as she now recalls,

> my father, he got killed in the mines, you know, and then they thought that maybe 'cause there were nine of us children, he [a company official] thought maybe he'd give me a job at the company store. So he let me work at the company store there and I used to stand on a little box 'cause the counter, you know, I was so little that I had to stand on a box so the people could see me and they changed me from there and I went up and worked in the soda fountain for a while.[18]

Company officials gave Mrs. Booker's brother a job as a trapper. Mrs. Booker related that the company did not have insurance on their employees when her father died, but that the following year they took out insurance on the men. Mrs. Booker did not receive any money from the company but "all the men worked that day [when] they had the funeral and they [each] gave a dollar, so [my mother] got $500."[19]

The company also aided some workers who suffered disabilities. Adolph Larson lost several fingers in a mining accident and subsequently was awarded around eight or nine hundred dollars from the company. The company had been operating a soft-drink factory, and at the time of Larson's accident they offered to sell the operation to him. Larson accepted the offer, putting about three or four hundred dollars down and paying the remainder of the debt later. The company purchased its soft drinks from Larson during the time that he operated the business, a period of four or five years.[20]

Consolidation's other role as the dominant company in a company town was that of community manager. Because company towns were not legally incorporated, companies like Consolidation operated in lieu of local governments. This meant that they could exercise total control over their employees' lives, both economically and socially, and the Consolidation Coal Company does seem to have exercised tight control over certain areas of the town, such as the company store. Former residents recall that the store was run in a highly efficient, orderly manner. In December 1907 the *Bystander* editor also noted that fact:

> The refined and cleanly appearance of the employees is extremely noticeable. In order to work in this store one must bring along his credentials and have a good record. The cigarette smoker can find no place here. If an employee is subject to strong drink or ever gets under its influence he is discharged. Clerks are required to respect themselves as well as their customers. Carelessness is not tolerated by Manager MacRae. He claims he has worked hard to make this a metropolitan store and he aims to keep it in that class.[21]

The *Bystander* editor also observed that the company would tolerate no boisterous or blasphemous language in the store. The editor concluded: "Mr. MacRae tells us that during his management of three years he has had cause only once to have a drunkard removed from the building."[22]

At the same time, however, Consolidation apparently exercised little if any control over the town's churches. This stands in sharp contrast to the experience of some company towns. In his study *Life, Work, and Rebellion in the Coal Fields*, David Corbin writes that company towns sometimes destroyed both the black church and the black preacher. Moreover, in company towns in southern West Virginia, the company supported only one church and one pastor.[23] In Pullman, Illinois, only one church existed. George Pullman, the town's founder, believed that it was wasteful to have a separate church for each denomination, so he constructed the Greenstone Church, intended to house all religious groups. In some company towns, companies fired employees if they did not attend church.[24]

The experience of both black and white churches in Buxton is almost diametrically opposed to these practices. Black churches flourished in Buxton, and according to reports in the *Bystander* by residents of Buxton, black congregations chose their own ministers and determined church policies. The two largest churches were the Mt.

Ebenezer Lutheran Church located in East Swede Town, a suburb of Buxton. (*Courtesy of Iowa Mines and Minerals Department, Des Moines*)

Olive Baptist Church and St. John's African Methodist Episcopal Church. Other black churches included a second Baptist church, known as the Tabernacle, and a Congregational church that met at the YMCA. It is not clear, however, whether the black churches leased land from the company or if the company donated the land. According to *Bystander* reports and private interviews, black ministers were highly regarded members of the community who, in addition to their religious duties, sometimes offered adult night-school courses, played active leadership roles in the YMCA, served as lodge officers, and frequently presented talks at school graduations and public events.

The white churches in Buxton, including the Ebenezer Lutheran, Slavic Lutheran, Methodist Episcopal, and Seventh-Day Adventist, also operated independently of Consolidation. Irene Goodwin recalled that for several years the white Methodist Episcopal Church had a student pastor from Iowa Wesleyan College at Mt. Pleasant. The student pastor had been assigned to the church by the Methodist district superintendent, who resided in southeastern Iowa (a practice followed in all communities). The first year, the student pastor commuted from Mt. Pleasant on weekends. The second year, he married and the church congregation rented a company house for his wife and him. The white Methodist Church was housed in the building that originally served as the Swedes' private elementary school.[25]

Consolidation also apparently exercised little influence over the professional people in Buxton, in effect allowing them to operate independently in their professional areas and in their leadership roles in the community. Again, this contrasts sharply with the situation in other parts of the country. Corbin notes that in southern West Virginia middle-class professional blacks like doctors, lawyers, and dentists were, "not looked upon as symbols of individual or racial progress, but as company 'licks' and spies, neither to be trusted nor emulated." In Buxton, George Woodson and Dr. E. A. Carter, both black men, occupied extremely prominent positions. George Woodson, as mentioned earlier, had earned a law degree at Howard University and had previously practiced law in Muchakinock, where he not only earned an excellent reputation as a lawyer but also played a major leadership role among Muchy's black population.[26] Of all the black professionals in Buxton, however, the most visible and widely acclaimed was Dr. E. A. Carter. Born in Charlottesville, Virginia, on April 11, 1881, Carter came to Muchakinock with his parents in 1882, where his father worked as a coal miner. His parents were ex-slaves. Carter graduated from eighth grade in Muchy in 1895 and from Oskaloosa High School in 1899. He then enrolled at the University of Iowa. Carter worked his way through college by waiting tables at local hotels and tending furnaces. In the summers he worked in the Buxton mines. After receiving his medical degree in 1907, he returned to Buxton and was employed by Dr. J. H. Henderson as assistant physician and surgeon for Consolidation. In 1910 he served as first assistant to Dr. Burke Powers, head physician for Consolidation, and later he was appointed to work with Dr. Ralph Early. Eventually Consolidation appointed Dr. Carter head physician. Dr. Carter played a prominent role in Buxton's social and educational life. In 1910 he was affiliated with the YMCA, serving as a member of two committees, Boys' Work and Education. Carter frequently spoke to school groups, church groups, and YMCA members. He remained in Buxton until 1919, when he moved to Detroit.[27] Many former Buxton residents remembered Dr. Carter both for his medical role and his exemplary personal qualities and talked in glowing terms about his many contributions to the Buxton community.

The view existed among almost all former residents that Consolidation treated its employees well. Earl Smith, who was never an employee of Consolidation but was a lifetime resident of the Buxton area, stated that "the company was awful good to [the employees] when they first come here." He pointed to the fact that the company had constructed the YMCAs and to the many activities that the "Ys" provided for local residents. He also recalled the company's construction of several big bandstands and its sponsorship of the Buxton Cornet Band. Smith related: "We used to, when we was kids, go down there with a horse and buggy and sit there in the buggy and listen to them play. The whole street was just full of people. There was no other place [to go to] in them days. . . . That was quite a recreation to hear that band play. They went all over the country playing too."[28] Other former residents also cited the YMCAs as an example of Consolidation's generosity.

While the perceptions of all former residents are important in assessing Consolidation's role as an employer, the attitudes of black former residents deserve special attention. The relationship Consolidation had with Buxton's black residents was different in all respects from the relationship it had with white residents. The great majority of blacks in Buxton resided there because they or their parents had been recruited by the company. Although blacks had originally been hired as strikebreakers and cheap laborers, after the founding of Buxton that situation changed. With the presence of the United Mine Workers, Consolidation had no choice but to pay black and white employees equal wages. It is instructive to note that Consolidation continued to hire blacks, in fact even to recruit them, after 1900 when the camp was unionized and, with standardized wages for all employees, the company no longer realized an economic savings by hiring blacks. It seems that the company perceived no difference between the work habits and productivity of blacks and whites.

Of considerable significance, moreover, are the views most blacks held toward the opportunities Consolidation offered them. For white miners who lived in Buxton, the community provided a good place to live; for most blacks who lived there, Buxton represented a near utopia. Through the offer of employment, they had been able to move out of the South, where most families had little material wealth and few, if any, industrial skills. In Buxton, black males earned good wages that provided their families with a comfortable standard of living. Hucey Hart remembered the prosperity of Buxton and the material comforts: "You never wanted for nothing. You go in there [the company store] if you want a suit of clothes. At that time they'd get a tailor-made suit of clothes made up for about thirty dollars, the best. You would go in there and say, 'I want a suit of clothes.' " Hart added that if employees wanted to borrow money, they only needed to speak to one of the cashiers and she would quickly lend the employee twenty-five or thirty dollars. He concluded: "You never wanted for nothing, your credit was good there."[29] Blacks also perceived Buxton housing to be excellent; many described the company houses as large or big. Their move to Buxton had raised them on the social and economic scale, which in effect made their lives more comfortable and their futures more optimistic. Blacks perceived that these changes and opportunities were made possible by the Consolidation Coal Company.

Once in Buxton, black residents discovered not only a higher standard of living than they had known before, but also a community that in their view was devoid of racial discrimination. Every black person interviewed stressed that he or she believed that Buxton was free from racial prejudice and that blacks felt no discrimination during their time there. Repeatedly, respondents stated that in Buxton everyone got along, everyone mixed together, and everyone attended social events at the main YMCA. In their perception, life there reflected total racial harmony.

Several black residents commented in almost poetic ways about their lives in Buxton. Dorothy Collier reflected: "But it was so much, you know, so much you could say . . . about Buxton. It could be a book within itself. A world within itself, you know." Gertrude Stokes commented: "I always said Buxton was like *Evangeline*. You remember in *Evangeline*, how they tore up that place, and I think Buxton was just on an order of *Evangeline*, those line of people. I always put them together. You know they were happy and somebody come along and destroyed them."[30]

Some blacks believed that the Consolidation Coal Company maintained this ideal setting by weeding out those elements that threatened racial harmony. As a black woman, Bessie Lewis remem-

The London family at home in Buxton. Taken around 1900, the formal portrait includes Minnie B. and W. H. London with their children, Vaeletta and Hubert. (*Courtesy of Vaeletta Fields*)

bered that the company, and particularly its superintendent, Ben Buxton, did a great deal to insure that blacks and whites were treated equally. Lewis recalled that Ben Buxton would not tolerate employees who did not treat black people well. She stated: "If he hired anybody that worked [but] didn't treat people right, he got rid of them. He didn't keep them, 'cause he believed in living and letting live and that was his motto."[31]

Lewis also commented on the way in which Buxton enforced his view that blacks should receive the same treatment as whites. She recalled that sometimes whites who had never lived around blacks found it "hard . . . 'cause they'd been used to doing everything they wanted to do and didn't give the colored people a chance. . . . Well, them kind of people . . . couldn't live there very long 'cause [the company would] fix it so they wouldn't be there." Lewis had known families who had lived in Buxton for five or six months but, she said, "they didn't have things the way they wanted them and they couldn't have it the way they wanted it, the way they had been living, so the next [thing] they'd be moving." Lewis believed that the company actively intervened to get rid of people who did not treat blacks in a fair and equitable manner, and she concluded that Buxton was a place that gave black people a chance. She remembered that the company had helped black young people attend school, learn different trades, and find jobs. She recalled one young man in particular whom the company had wanted to employ in an office, so "they'd send him to school somewhere and get training and come back, [and] pick up a stenographer's job."[32]

The presence and purpose of the YMCAs also deserve attention in regard to the relationship between black residents and the company. Although both races used the "Y" facilities, the insititutions were clearly developed as recreational and educational facilities for blacks. That fact seems to indicate that the company anticipated continuing its policy of hiring southern blacks. In establishing the Buxton YMCAs, Consolidation was, no doubt, genuinely interested in promoting a satisfied, contented work force, but as previously discussed, they may also have been influenced by the welfare capitalism movement. In his study of the coal industry in southern West Virginia, David Corbin notes that, beginning in the mid-teens, coal operators there built YMCAs for their employees not simply for their amusement, "but to create a 'sentiment' among the miners." Operators hoped that the YMCAs would bring about reconciliation between the coal companies and the miners. One coal operator stated: "There has always been more difficulty in getting employees and employers together for the purpose of discussing a difference than there ever was in finally settling it. Neither of the parties in a controversy can approach a problem in an intelligent manner if they have been denouncing or endeavoring to take an unfair advantage of one another." At the same time, YMCA officials stressed that the presence of YMCAs encouraged the miners to use their leisure time wisely. In 1914 a YMCA official stated before a group of West Virginia coal officials that "no man is likely to become dangerous to society as long as he is working. It is misuse of leisure that in the final analysis works ruin to our laboring men." The official added that, in turn, the misuse of leisure time led to labor agitation and the use of alcohol.[33]

While it is difficult to know how the YMCAs at Buxton compared in size and facilities with those in other coalfields, it is known that the Buxton "Y" facilities were extensive. The "Ys" offered males the opportunity to swim; play tennis, basketball, or billiards; roller skate; or attend movies, dances, and various lodge meetings. According to the *Bystander*, the YMCAs offered a wide range of activities at all times for younger boys, young adults, and adult men. For many years the YMCAs hired both a general director and a director of the boys' department. At times the general director taught night courses. Certainly the presence of the two "Ys" helped eliminate the "misuse of leisure time." At the same time, Consolidation officials prohibited the sale of liquor at the company store or at any business on company property.[34] Undoubtedly, the presence of the YMCAs helped promote the company's strongly held view that miners should not consume alcoholic beverages. The *Bystander* routinely reported on activities at the YMCAs, as typified by the following:

> The YMCA has entered into its fall and winter work and the men and boys are spending many pleasant hours each day in the two buildings. Sunday afternoon at 3:30 there will be the regular afternoon meeting. Be sure that you are there to hear the speaker of the afternoon. Tuesday night is Bible class night. Come out and bring a friend. A glee club has been formed and the men are doing fine. . . . The different gym classes have been started for both men and boys. Football and basketball practice has also started and the men and boys are taking [a] great deal of interest in these two sports, as well as indoor baseball, handball and many other games. Many are finding enjoyment in our fine swimming pool.[35]

Black men also organized other groups at the YMCA, as well as sponsoring individual events and assisting outside groups in promoting programs of mutual interest. In February 1906 the young men formed the YMCA Literary Society. In the early years of Buxton, male residents formed the Buxton Christian Workers' Alliance, which in 1908 sent delegates to the state temperance convention in Des Moines. Apparently the group disbanded after a time, however, because in January 1911 the town's ministers and deacons were again meeting to form a Christian Workers' Alliance. YMCA officials and prominent members strongly supported the Alliance's temperance stand. The young men also sponsored many purely social activities. Marjorie Brown recalled that "the young men had a social club. You went to their balls by invitation, and everybody that was anybody went to that ball, with a tailor-made suit and some of the most beautiful gowns I ever saw."[36]

Many black former residents remembered the important role that the YMCAs played in their lives. Jacob Brown belonged to the "Y" when he lived in Buxton, and he particularly liked to visit the reading room. Brown remembered that "they had history books there and . . . well, they got all kinds of magazines and things like that and newspapers." Brown specifically remembered that the "Y" carried newspapers from Pennsylvania cities and from Chicago, New York, and Des Moines. Mike Onder related that the reading room also contained checkerboards. Onder described the room as resembling a hotel lobby, with many big chairs scattered around. Brown also enjoyed using the gym facilities, and he remembered that movies were shown there on Monday, Wednesday, Friday, and Saturday nights.[37]

Leroy Wright also remembered the "Ys" in a positive manner. He stated that as a young boy he swam in the heated pool in the little "Y." As a young man he learned to play pool in Buxton, although no males were allowed in the main "Y" (which contained the pool tables) until they were twenty-one. Wright remembered that by the age of sixteen he had become an excellent pool player. When the main "Y" officials learned about Wright's proficiency in the game, they allowed him to play there. Moreover, Wright learned to play a musical instrument at the main "Y," and he also related that a Mr. Potter held Bible classes there every Tuesday night. Harold Reasby recalled that many road shows played in the auditorium. Some were all-black musical shows in which the cast traveled to Buxton in Pullman cars and remained for an entire month. Although not mentioned by any informant, the Boys' Department of the YMCA organized a Boy Scout troop in 1913.[38]

It is not clear to what extent white residents used the facilities of the YMCAs. Members of the United Mine Workers of America held their meetings in the large "Y," so in that instance both black and white members attended. Interviews with white former residents give a somewhat mixed view of their use of "Y" facilities. Some white informants recalled attending movies and dances at the YMCAs and also roller skating there. Some remembered the "Y" as the social center for both white and black residents. Other informants viewed the "Y" as primarily the blacks' territory, and they did not feel particularly welcome there. Some recalled that they expressly avoided the "Y" because some black youths there were not very friendly to whites. Yet, even though white Buxton residents did not frequent the YMCAs as often as blacks, interviews and *Bystander* accounts indicate that the YMCAs played a vital role in the lives of all Buxton residents. They offered social opportunities and physical facilities that were unknown in all but the largest Iowa cities during the first two decades of the twentieth century. Even if whites did not use the gym or other physical

The Buxton YMCA basketball team in 1907. Herman Brooks is in the front row, left. (*Courtesy of Herman Brooks*)

facilities often, they did attend social, educational, and cultural presentations there.

For black residents of Buxton, the presence of the YMCAs must have produced important emotional results, as well as more visible intellectual and physical benefits. Of great importance was the fact that they were staffed and controlled by black males. As far as can be determined, directors of both the main "Y" and the Boys' Department were always black men. "Y" committees were composed of black Buxton residents. Moreover, according to *Bystander* accounts, most speakers (aside from group presentations) were black men. Within these two black institutions, young black males had the opportunity to interact on a personal basis with local black community leaders, including many professionals. Further, the main YMCA provided black leaders with a forum from which they could speak to the black male membership on a variety of topics, many of which were motivational in nature. Black professional men, local residents, and outsiders alike continually urged young blacks to further their education, save their money, and abstain from the use of alcohol. These same men frequently talked about the need for blacks to achieve better occupational positions and to live moral lives, all of which, they said, would reflect positively on the black race. At least tacitly, they conveyed the message: Take pride in being black.

Through the very presence of local professional men like Dr. Carter, George Woodson, and S. Joe Brown, many young black males had successful role models for the first time. Most black youngsters came from homes where their parents had only a limited education and where their fathers labored as semiskilled workers. Even though many of these same parents stressed hard work and the value of education, the main YMCA, with its aura of professionalism and respectability, provided an exposure to a world of success and possibly conveyed to the young men that higher education and professional status were not beyond the limits of their abilities. Certainly the presence of successful black men appearing within an institution totally controlled by blacks must have instilled a sense of dignity and worth in the young men involved. The following statement written by a black minister, Rev. H. W. Porter, who traveled from Des Moines to speak at the main "Y," illustrates that point:

> Now the grandest of all was the scene at the YMCA Hall. The building is the neatest and best equipped I have visited. The reading room with the latest periodicals, and growing flowers is especially inviting; everything is done to

make the men feel at home, and give to them comforts which can only be known, as we see the well kept building and note the progress of the work. At 4 p.m. we entered the auditorium and as the music began it was a sight to behold, and one to be appreciated, to see 200 men marching into the auditorium. The singing was good, and we have never spoken to a more intelligent and appreciative audience—men whose faces beamed with a desire for knowledge.[39]

In reporting the activities of the YMCAs, the *Bystander* frequently lauded Carter, Woodson, and other black leaders for providing positive examples for Buxton's young black males. Between 1900 and 1915, it published many articles and editorials that referred to the need for prominent Buxton blacks to provide leadership for younger members of the community and to support their fellow blacks in economic, moral, and political endeavors. The *Bystander* referred to men who acted in this manner as "real race men." The newspaper cited Dr. Carter, George Woodson, B. F. Cooper, and Reuben Gaines, Sr., as being "real race men."

At the same time, the *Bystander* singled out black men who were not supportive of their race. In 1918 the *Bystander* editor made his annual trip, and in his subsequent report he singled out several black men who had done well in business. The editor then added:

Our race has some men in it that are running from their own race, a disgrace to both races. Yet we are sorry to say that the richest negro in this county, who owns over 1,200 acres of the best land in the county, and as the county treasurer of this, Monroe County told me, this man's land was the highest assessed of any land in the county, and yet this is as described above, no benefit to his race. Standing as he does, he could have been a wonderful help to our struggling race, but he is a deserter, a slacker.[40]

Although the *Bystander* editor did not name the "slacker," there is no doubt that he was describing Hobe Armstrong. Armstrong clearly had experienced the greatest business success of any black in the area and was the largest landowner. Although the *Bystander* made no mention of Armstrong's strong affiliation with the company, it was a well-known fact to everyone in and around Buxton.

Even with prosperous economic conditions and a pleasant physical environment, Consolidation had some critics. While most miners and their families believed Consolidation to be a generous and concerned employer, at least a few workers viewed the whole process as strictly economic, where the company rarely, if ever, acted in a generous manner. Moreover, at least one employee believed that the company exercised almost total control over its employees' lives, which always resulted in an advantage for the company. Robert Wheels remembered vividly that in regard to buying at the company store, "th company treated you so nice, you know, so fair. You could get anything you wanted, that is, anything, you understand. And when you got down [in debt] in there, well, you were just there." Although a the time most black residents believed they had control over their own lives, Wheels asserted that in reality they did not. The company totally controlled the town, the jobs, the store, and all other activities "The company," he said, "was responsible for everything, that is, anything you wanted. You take undertaker, doctors, or anything, they had them and they were responsible for it." Wheels explained that he did not realize until after he left Buxton that his life there had been so controlled by others. He acknowledged that Buxton was a good experience in many ways: "I'm glad of that. . . . We had what we wanted, as much as we wanted, and everything, . . . and still you was a slave in a way of speaking." After brief reflection, Wheels added: "That is, you take back in them days, you thought it was [free] but it wasn't. Everything the company, every little inch they put on, you paid for it, everything." After leaving Buxton, Wheels worked for thirty-one years for the Rock Island Railroad. He concluded that he was glad he had left Buxton, because otherwise "I wouldn't have known how to do nothing but coal mine, or get out and work a little bit on a farm or something like that. The work here [in Des Moines] was an education for me."[41]

Buxton newspapers also provide evidence that not all Buxton residents approved of company policies. During its existence Buxton contained a variety of newspapers, all of which existed for only a short time. Only a few issues of the *Buxton Advocate*, the *Buxton Gazette*, the *Buxton Breeze*, and the *Buxton Eagle* are known to exist, so it is impossible to know what consistent editorial positions, if any, the newspapers expressed. One editorial has survived, however, from the *Buxton Eagle* in which the editor took strong exception to certain policies of the company. On April 27, 1905, the *Eagle* published an editorial entitled "As to Saloons in Buxton." Although the editor expressed strong opposition to the presence of saloons and the sale of whiskey, he also added a strong condemnation of Consolidation for several of its policies. The editor wrote:

And the very fact that an unscrupulous corporation can huddle six thousand laboring people together in a camp away from communication and facilities to conveniently reach the outside world, keep them under the thumb of one man, dependent upon his wish for water, fuel, shelter, food, clothes, light, air, education, religion, politics and a burying ground, is contrary to the spirit of our laws and free institutions. The corporation's practice of sending their secret agents stealthily into the eastern and southern states to smuggle any kind of men that will accept the transportation and come by car loads into one of the freest and best counties in Iowa, is an immoral and unlawful proposition that must sooner or later engage the serious attention of the law makers of Iowa.

This corporation can and does not only menace the peace and mar the progress of the people of Buxton by refusing to incorporate the town, but it endangers the welfare of the whole county by meddling with and dictating the political affairs of the whole people.[42]

The editor also condemned the company for having "ruined the educational principle of the town by establishing and maintaining a separate white school and at the same time placing a white principal over the deserving and cultured colored teachers of the colored school."[43] It is not known how the company reacted to the *Eagle*'s editorial. It is known that the *Eagle* only operated as a newspaper for several years. Given the monolithic power of Consolidation, however, it is difficult to imagine that company officials tolerated criticism kindly. But, in general, critics of the company seem to have been extremely limited in number.

While questions remain as to the motivation of Consolidation officials in developing a progressive, egalitarian community, there is little doubt as to the success of that development. On all counts, Consolidation proved itself to be a liberal employer, held up for recognition by employees, fellow coal operators, UMW personnel, and state mine officials alike. Within the state, Consolidation clearly maintained equitable hiring policies, provided modern and comfortable housing, and maintained a pleasant physical environment. Consolidation provided these conditions in spite of the fact that its employees had joined the UMW. In other words, unlike so many coal companies, Buxton's improved living and working conditions were not intended to prevent employees from unionizing. These actions indicate that Consolidation was an atypical coal company, both for Iowa and the nation.

In attempting to determine Consolidation's motivation for establishing Buxton as a planned model community, there seem to be two alternatives. The first rests with the Buxton family, father and son, who controlled the company for almost thirty years. Because there is little extant information about these men, it is not possible to know if they shared a belief in racial equality or if they desired somehow to provide social and economic assistance to southern blacks. The second, and probably more likely, alternative rests on the set of practices known as welfare capitalism. Here two explanations are possible. First, it is likely that Consolidation officials, including John and Ben Buxton, were at least indirectly influenced by the ideas of welfare capitalism, certainly having heard about the practices of Pullman and others. Second, it is also possible that Consolidation officials, through direct contact with other industrialists, decided to pursue such a policy. Consolidation had a board of directors that met in Chicago on a regular basis. Nothing is known of these men except their names, but it is entirely possible that these directors had firsthand knowledge of the activities of welfare capitalists and decided to emulate these practices at their Iowa property. Regardless of these possible alternatives, however, one thing is certain: the Consolidation Coal Company did pursue policies that at least paralleled welfare capitalism, and because of that fact they provided Buxton residents with a quality of life far superior to that found in other Iowa coal camps.

5

FAMILY LIFE

INDIVIDUALS arriving in Buxton in the early twentieth century would find a community where family life dominated all social and economic activities. Most residents traveled to Buxton in family units, and once there, like coal mining populations elsewhere at the time, the great majority of residents lived in household units. As a result, the family served as the basic unit of social organization, and social and economic activities stemmed out of that organization. In most homes, men went to work in the local mines, while women remained at home to manage the household and care for the children. Parents raised their children in a closely supervised manner, and in many homes children began to contribute to the family income at an early age. Because children worked both in and outside of the home, the family existed as both an economic and a social unit. Family members, moreover, assisted one another in times of trouble. Both black and white families exhibited stability and solidarity; throughout the Buxton experience, kinship ties remained strong. For blacks and whites alike, the family stood at the center of people's life and work.

Within the family setting, both black and white women played a wide variety of social and economic roles. Interviews and census data show that, in keeping with the social norms of the early twentieth century, most married women in Buxton did not seek employment outside the home, and interviews with former residents indicate that most male and female roles were clearly delineated. Married women carried out all domestic tasks, including housecleaning, food preservation, meal preparation, laundering, and child care. Because most families in Buxton had house lots of at least a quarter acre, families combined the characteristics of town and country living. Most homes were productive units in which women produced a major portion of the food and goods consumed by their families. Overall, the type and extent of domestic work did not vary significantly between black and white households. The major consideration in determining domestic work was family size. At the same time, all family activity was affected by the fact that Buxton was a company town and that the dominant occupation was coal mining.[1]

Interviews with former Buxton residents indicated that most women had a well-defined household routine. Many women rose at 5 A.M. to begin cooking breakfast and to prepare their husbands' lunch. Coal mining families had to rise early because the men boarded the miners' train at 6 A.M. Jacob Brown recalled that he started the fire in the kitchen range and his wife cooked breakfast. Breakfast in the Brown household usually consisted of freshly baked biscuits and sometimes beefsteak or pork chops. Mrs. Brown packed some of the leftover meat in Brown's lunch bucket for his noon meal in the mines. She usually packed three sandwiches—two meat and one jelly—and a piece of cake or pie.[2] If the family kept boarders, the housewife also had to prepare breakfast for these men and pack their lunches. In many families one or two sons also worked in the mines. Bessie Lewis related that at one time four of her brothers, as well as her father, worked as coal miners. Moreover, Bessie's family sometimes kept boarders. Her mother then had at least five miners' lunch buckets to pack every morning.[3] In many homes the housewife also had to prepare the children's school lunches. Some homemakers had a dozen lunch buckets to fill daily.

Although most Buxton women had considerable flexibility in scheduling their daily activities, these activities had to accommodate the men's return at the end of the workday. Sometimes the miners' train arrived as early as 4 P.M. and sometimes not until 6 P.M. The men left home in the morning wearing their miners' clothing and returned home wearing the same clothing. Consolidation's mines did not contain showers, so once off the train the men usually headed home to bathe and change clothes. The routine in most mining households required that the women have hot water ready for their husbands' bath as soon as they arrived home and that the evening meal be served immediately after they finished bathing. Women usually put the tub in the kitchen close to the cookstove. Gertrude Stokes remembered that women in Buxton often went out during the day to shop and socialize, but that they always had to be back home with the hot bathwater waiting when their husbands arrived.[4]

Given the men's need to bath daily and the sometimes cramped conditions in Buxton homes, there was often little privacy for any family member. Some families did have facilities, however, so that bathing could be a private matter. Lara Wardelin remembered that in her family there was a great concern for privacy. She recalled, "Way back then people were very, very protective for their children. I never did see my father take a bath. They had these little outside houses where they'd take their baths and things. They never did take a bath in front of us." Kathren Brown heated an upstairs bedroom where she pulled out an old tub for bathing purposes.[5]

Throughout the day, Buxton women performed all the household tasks common to homemakers in the first quarter of the twentieth century. Unlike their counterparts in cities and incorporated towns, Buxton women did not have the use of electricity. Therefore the women did not enjoy such laborsaving devices as electric washing machines, vacuum cleaners, electric kitchen ranges, or electric lights. Food preparation and preservation seemed to occupy the most time. Women, along with their husbands and children, raised large gardens, which required constant care during the summer months. Some families also cultivated fruit trees. Canning and drying were the major means of food preservation. Most families supplied some of their own meat through butchering; many families butchered a hog at least once a year. Butchering was apparently a family affair, with both the housewife and her husband doing part of the work. If the family kept a cow, they also produced their own milk. A few informants remembered that their mothers churned butter. Families living on the outskirts of Buxton or at the end of streets often kept chickens, so they had their own supply of eggs and poultry meat. Overall, many families in Buxton were largely self-sufficient in food.

In addition to raising and preserving food, women also faced the endless task of meal preparation. On a typical day, housewives prepared two major meals, breakfast and supper. Both meals typically contained meat and potatoes. The evening meal also contained vegetables and frequently fresh or canned fruit and possibly baked goods. Bread was an important item, both for home meals and for lunches. Given the large families, most women must have baked bread several times each week. Agnes Erickson remembered that her mother baked bread three or four times a week. Women also frequently baked pies, cakes, and cookies. Dorothy Collier related that her mother always did

Company houses, near Buxton's main business district. Note the gardens planted alongside or behind the houses. The first company store is located in the middle background and the two YMCAs are at its left. (*Courtesy of Iowa State Historical Department*)

a great deal of baking and general food preparation on Saturday. Men in the mines did heavy physical work and therefore required large meals at home as well as substantial noon lunches.[6]

While most domestic and child-rearing tasks remained the same in black and white households, an important consideration for women in regard to household work was family size. According to interviews with former Buxton residents, numerous black families had nine or more children. Mattie Murray was one of seventeen children, while Robert Wheels was one of fifteen. Oliver Burkett's mother had eleven children. Bessie Lewis's mother had ten children, while the mothers of Odessa Booker and Hucey Hart each had nine children. Susie Robinson experienced seventeen pregnancies, although she did not carry all fetuses to full term. For these women, all household duties became more onerous, and the hours spent doing laundry, sewing, and cooking were obviously far greater than for women with small or medium-sized families.[7]

Women with large families found washday to be particularly demanding. Washday involved heating water in large boilers, pouring it into a washtub, and then scrubbing the clothes on a washboard. Most families kept several rain barrels to catch soft rainwater, and they also had cisterns that provided them with water. Washing machines powered with gasoline motors became available in the early 1900s, but no informants mentioned their mothers having such machines. Washday was made particularly difficult because of the miners' clothing. Pit pants were made of heavy canvas material and were worn until they were filthy with dirt and coal dust, and each man used several pairs of pit pants each week. White clothes required extra work, as they first had to be boiled.[8]

Many women in Buxton sewed most of their family's clothing. Mattie Murray remembered that her mother had no choice, with seventeen children in the family. Mrs. Murray stated that many times her father went to collect his wages and found that he had no money left after deductions, so her mother would go to the store and purchase bolts of cloth rather than buying ready-made clothing. She remembered that "all the girls had dresses alike. The boys had shirts alike."[9] When asked how her mother found time to sew while caring for seventeen children, she replied:

> Mamma used to set up nights. On Saturday night she'd set up all night long so we'd have new things to wear to Sunday school. Make the boys shirts and us

girls dresses and she'd take flour sacks and make us underclothes out of them. The man down to the company store, he'd save her all the flour sacks that he'd have. You know, they'd put the flour in big barrels and he'd shake the sacks out and stack them up and save these for Sister Bradley. And Mama would go down there and she'd have a stack that high of flour sacks. When the wind would blow I'd have More Pillsbury's Best across my pants.[10]

Many women found their domestic work to be extremely taxing. Odessa Booker related that her mother rarely went out of the home, because of her extremely heavy domestic work load. She stated: "I remember that she'd be so tired at night she'd sit up in a chair and sleep, and I used to just slip her shoes off and she said 'let me sit here a minute' and after that she'd just lay across the bed, the foot of the bed, 'cause she had to get up early and get Daddy's lunch, fix his bucket, you know." Other times, Mrs. Booker's mother would fall asleep in a chair. Mrs. Booker recalled: "Sometimes I'd say 'Mama why don't you go lay down' and I'd pull off her shoes . . . and she had a great big chair then. She'd sit in that chair and sleep most of the time." Bessie Lewis remembered that her mother worked exceedingly hard too. When asked about her mother's routine, she explained: "Well, she had to clean up and help sew for the kids, you know, and do her housework and things around. She had plenty to do. The kids grew up [and] she had to teach them how to do. . . . A lot of cooking to do with a large family, you know." Oliver Burkett related that his mother, Mary Elizabeth, worked extremely hard raising a family of eleven children. When asked if his mother belonged to any organizations outside the home, Burkett responded: "Mama, she didn't have time to do nothing but raise us." He added: "She died an early death."[11]

An additional health concern for women was the matter of pregnancy. According to our interviewees, many women used the services of midwives for delivery and postnatal care. Although existing Monroe County birth records do not allow for a complete tabulation of Buxton women who used the services of midwives, partial birth records have survived. In 1904 and 1905 in Bluff Creek Township, thirty-five white women used physicians for childbirth, while four used midwives. For black women in the same period, eighteen were attended by physicians, while fifty used midwives. According to this limited data, far more black women preferred the services of midwives than did white women.[12]

Buxton had both black and white midwives. Several informants remembered Lucy Mealy, a black woman who delivered both black and white babies. Jacob Brown recalled that many women in Buxton used her services: "She'd come to your house about a week ahead of time. She always carried a black bag. . . . All us kids, we'd say, 'She's going to bring a baby,' you know. 'Cause at that time they didn't talk about it like you do now." Margaret Rhodes was another black woman in Buxton who practiced midwifery.[13] Agnes Erickson recalled that East and West Swede Town each had its own Swedish midwives. Erickson stated that in the early years many women in Buxton used midwives, but eventually many women began going to the hospital for delivery. Irene Goodwin recalled that her grandmother, Mary Chambers, worked as a midwife in both Muchakinock and Buxton for many years. In total, Mrs. Chambers delivered about a thousand babies. Mrs. Chambers, a white woman and the wife of a coal miner, delivered babies mostly for white women in the neighborhood, but her granddaughter recalled at least a few instances where Mrs. Chambers delivered black and Indian babies.[14]

The reasons why women used midwives apparently varied. Susie Robinson recalled that doctors "didn't fool with nothing like that," so it was necessary to use midwives. Some women simply felt more comfortable with female midwives than with male doctors. Mattie Murray claimed that practically all of the women in Buxton used midwives, because "They didn't like doctors. . . . They didn't feel comfortable goin' with him so they had old ladies . . . [who] would bring the babies into the world." It is also likely that many women, foreign-born and native-born alike, had experienced the use of midwives in their former homes and therefore carried the tradition along to Iowa. The most common reason, however, was that women liked the postnatal care that midwives provided. Susie Robinson explained that midwives were "much better for you, too, because they'd take care of your kid for nine days, wash him up and everything and clean him up for you . . . and show you how to keep him, you know." Several former residents recalled that the midwife charged five dollars for all services associated with the birth of one infant. The use of midwives did have an additional cost, however. While the company medical plan covered the cost of delivery by the company physician, it did not cover the cost of a midwife's services.[15]

Testimony from the daughters of Buxton midwives supports the other interview data in regard to the care provided by midwives. Bessie Lewis recalled that her mother attended the women before the birth as well as after. Mrs. Lewis visited mother and child for two weeks after the birth and bathed the baby each day. According to Irene Goodwin, her grandmother Mary Chambers went into the patient's home and remained there for the full nine days. She did not do housework or cooking, but she did provide complete care for the mother and infant. She received ten dollars for her services.[16]

Several women remembered the postnatal practices prescribed by both the midwives and the company physicians. The first rule was that women remain in bed for nine days following delivery. Susie Robinson explained that women were not allowed to go outside for a full month after delivering. When Mrs. Robinson did get out "the ground looked like it's been raised up and the grass and everything look funny." She added that new mothers also had extremely restricted diets: "I guess people now eat pork chops, eat all kinds of stuff when they're down. In those days you couldn't eat nothing but crackers and cheese."[17]

In addition to the help of midwives, some women were also assisted by family members and friends. During her confinements, Susie Robinson's younger sister "was always with me. . . . Every kid I had, she'd come. Don't care where she was at, she'd come. If she was out of town, she'd come stay with me until I got so I could take care of the kids." Odessa Booker remembered that neighbors came in and helped her mother at these times: "I remember the lady next door used to come in when the baby was small, you know, and then of course when the baby got big enough I used to lay it on the table and wipe it. Mama showed me how to do [that]."[18]

While large families represented heavy work loads for women, most children were expected to help with household tasks from a fairly early age. In this sense, children had well-defined roles. Both black and white informants stressed that during their preadolescent years their parents expected them to accept considerable responsibility and perform a variety of household chores. Girls primarily assisted their mothers within the home, and some children in large families had to accept major responsibility at an early age. Odessa Booker, the oldest child in a family of nine, had to assist her mother in many domestic tasks as well as caring for her younger brothers and sisters. "I was the oldest," she said, "and I had to do all the work. . . . I couldn't play 'cause I had to watch the baby while Mama cooked. [The others were] out playing and I had to see that the [younger] kids got dressed. I was the next mama." Mrs. Booker added: "I washed dishes and . . . every

Saturday we had to scrub floors and you know they had shelves then. You had to dust the shelves and wash the lamp globes, the lamp chimneys, and fill all the lamps up. . . . And you had to do all of that, set the table, wash the dishes, feed the babies." Mrs. Booker and her brothers chopped wood and brought in nine buckets of coal each day. The family had two stoves in the house, and the nine bucketsful provided one day's supply of fuel. Mrs. Booker also helped with the laundry. Her mother did the family's wash on a washboard, and Odessa helped wring out all the big items of clothing and blankets.[19] Wilma Stewart, a member of a Swedish family, lived on a farm near Buxton. She remembered that as a young girl she helped prepare breakfast and do dishes before going to school. She did not have to help much with outside work, but she did gather eggs daily. Mrs. Stewart also helped her mother with the washing and ironing.[20]

Many other people who spent their childhood in Buxton also recalled having after-school responsibilities. Hazel Stapleton remembered that her brothers had to bring in coal and chop wood. The Stapleton family lived on an acreage near Buxton. The family kept chickens, and it was the children's responsibility to gather the eggs. Hazel's oldest brother helped their father clean the barn. After school, Charles Lenger and his sister and brother always had chores to do. They had to fill the coal buckets, chop and bring in wood, and fill the water containers. Every night Hucey Hart had to fill ten or fifteen small cans with pieces of coal, chop kindling wood, and draw water from the cistern.[21] Before going to school, Earl Smith sometimes helped with the milking, and after school he brought in coal for the night. Many informants stated that during the summer they worked in the family garden.[22]

In addition to domestic responsibilities, many black women also performed work for money, thus adding to the family's total income. Most of these jobs were offshoots of the women's roles as homemakers. Interviews indicate that married women, often with the help of their daughters, worked both in and outside of the home to earn substantial amounts of money. Helen Duke recalled that her mother, Mrs. John L. Lewis, was self-employed most of the time. "She cooked a lot. She was a seamstress. She was just a person that could do anything. She crocheted. She sewed, you know, and at that time there in Buxton she did most of the sewing for a lot of people." Before Buxton, Mrs. Lewis had lived in Albia, where her father had hired a seamstress to come into the home to teach his daughter to sew and crochet.

Lester Beaman's mother took in washing during the time the family lived in Buxton. Mrs. Beaman did large washes, mainly for white families who lived in town. Beaman first had to carry the water and then the wood to heat the water. Later he and his brother and sister delivered the laundered clothes to her customers.[23]

Other black married women earned money both by working outside the home and by selling food products. For a time, Alice Neal worked as a cook. During the week Mrs. Neal lived in Miami, which was also known as Number 10 Junction and was located about ten miles west of Buxton. Her two children stayed with their grandmother in Buxton during the week, and on weekends they visited their mother in Miami. The daughter, Dorothy Collier, remembered that as a little girl it was a great thrill to ride in the engine of the train from Buxton to Miami. She also remembered that an aunt who lived a few miles outside of Buxton sold butter and eggs to the company store. As a young girl, Mrs. Collier rode horseback to bring the butter in to the store. She remembered: "[my aunt] had a phone. She'd call and the [store clerks] would meet me at the door and take me off of the horse." Odessa Booker's mother raised mustard greens in her yard for sale to local people. Mrs. Booker remembered: "We'd pick them and wash them and get a basketful for a quarter. We kids did that, you know." Her mother also sold beans and corn, which she raised in the family garden.[24]

Some black married women contributed to the family finances indirectly. In other words, they did work that eventually led to a higher family income, although they themselves did not receive a specific wage. Gertrude Stokes's father operated a livery stable and her mother drove a carriage providing passenger service between Buxton and Albia. Stokes related that her mother "used to carry passengers places. . . . We had a horse and two-seated carriage. She used to carry passengers backward and forward from Albia, backward and forward wherever they wanted to go."[25]

The major way that married women in Buxton earned money, however, was by taking in boarders. For some women it seems to have been strictly a matter of economics: They took in boarders simply to bring more money into the household. In these cases, boarders were sometimes of a different race or ethnic background from the family itself. Nellie Lash King recalled that her mother frequently kept black boarders, although the Lash family was white. For other women, ethnic considerations were involved. Charles Lenger recalled that "men coming from the old country didn't have a place to stay, and so they asked if they could stay at your home, and they'd pay their board and all that." No doubt Lenger's mother was responding at least partially out of a sense of duty in helping young men of the same nationality as her family. Alex and Agnes Erickson, both second-generation Swedish-Americans, remembered that their mother only took in boarders from Sweden: "We had four and two, two and four. Not all the time, but they'd come from Sweden and they could talk Swede to us, and when they'd come someone who could talk American generally would take them with them to work in the mine 'til they learned to work and learned to talk so they could handle themselves." The boarders lived in one bedroom in the Erickson house.[26]

Boarding practices in Buxton approximated those in other coal mining communities in Iowa. While most camps contained at least one boarding house, many single miners preferred to live within a family setting and thus preferred private boarding arrangements. In many homes, the boarders were treated like family members. Boarding practices were not restricted to any particular ethnic group but seemed to be equally prevalent among all groups, including blacks. In Iowa's coal communities, Italian-Americans comprised the largest foreign-born group after 1900. In countless Iowa coal mining communities, like Seymour and Numa, first-generation Italian-American women took in single Italian men as boarders both to earn money and to discharge a sense of responsibility that they felt for them, since they were so far away from their own families.[27]

The boarding rates charged by Buxton women varied considerably. Josephine Erickson charged about fifty cents a day. Bessie Lewis's mother earned considerably more, charging each boarder $7.50 a week. For these fees, women provided the boarders with a place to sleep, meals, and laundry service.[28] Italian-American women elsewhere in Iowa's coal camps charged two dollars a week for all services, including food. The number of boarders within those households varied from one to eight, with the average number being three or four. Bodnar, Simon, and Weber found that in Pittsburgh blacks, Poles, and Italians charged boarders three to four dollars a month, plus the cost of food. They estimated that total boarding rates for an individual averaged eleven dollars a month. Approximately one-third of the families took in an average of three to five boarders and thus earned about twelve to sixteen dollars a month from boarding fees. In contrast, the amount earned each month by Buxton women, particularly if they took in from three to five boarders, was considerably higher than that earned by women in other Iowa camps or in Pittsburgh.[29]

While interviews with former residents provide one means of identifying occupations for women, a second method is through occupations listed in the census. According to the state censuses of 1905 and 1915, only a small proportion of Buxton women listed occupations in either census. Tables 5.1 and 5.2 show that black females constituted 25 percent of the total population of Buxton in 1905 and only 10 percent of them listed occupations. Table 5.3 shows that the two major occupations listed for black women in 1905 were domestic and housekeeper, jobs that required little, if any, formal education. Twenty-one black women listed dressmaking, which did require special training. Nine listed "boarding house" as their occupation, presumably meaning that they operated boarding houses. Seven black women listed teacher. Presumably both black and white teachers had some professional training.[30]

In 1915 the total number of black females in Buxton had decreased by roughly a third, and the number of black women listing occupations had risen slightly to 11 percent (Tables 5.1 and 5.2). Table 5.4 shows that black women did experience some changes in occupations. The number of domestics had risen considerably, from twenty-nine in 1905 to forty-eight in 1915. The second major occupation for black women in 1915 was teaching, with six women teaching in the

Table 5.1. Female Racial Composition, Buxton, 1905 and 1915

Race	1905	Percent of Total Population	1915	Percent of Total Population
White	943	19	1220	27
Black	1254	25	836	18
Total	2197	44	2056	45

Source: Iowa, Census of Iowa, 1905 and 1915, Manuscript Population Schedules for Bluff Creek Township, Monroe County.

Table 5.2. Females Listing Occupations, Buxton, 1905 and 1915

	White		Black	
Year	Number	Percent	Number	Percent
1905	105	11	124	10
1915	79	6	91	11

Source: Iowa, Census of Iowa, 1905 and 1915, Manuscript Population Schedules for Bluff Creek Township, Monroe County.

Table 5.3. Occupations of Females, Buxton, 1905

Occupation	White	Black
Laborer	1	—
Clerk	3	2
Farmer	14	1
Teacher	12	7
Insurance agent	1	—
Bookkeeper	4	—
Postmaster	—	1
Milliner	1	—
Dressmaker	7	21
Music teacher	1	2
Wash woman	—	2
Domestic	23	29
Agent	—	1
Nurse	1	—
Cook	1	1
Waitress	1	—
Housekeeper	9	29
Stenographer	1	1
Bank cashier	1	—
Canvassing	2	—
Manager, lumberyard	—	1
Sewing/boarding	1	—
Boarding house	4	9
Hairdresser	—	1
Midwife	—	1
General secretary, YMCA	1	—
Wine maker	1	—
Post office clerk	—	1
Other	12	13
Total	102	123

Source: Iowa, Census of Iowa, 1905, Manuscript Population Schedules for Bluff Creek Township, Monroe County.

public school and one woman teaching music. Dressmaking ranked third.[31]

Occupations listed by white women in 1905 and 1915 differed somewhat from those of black women. Tables 5.1 and 5.2 show that while white females accounted for 19 percent of the total population of Buxton in 1905, 11 percent of the total number of white females listed occupations. The major occupation was domestic. The occupation of farmer ranked second and teacher third. Housekeeper and dressmaker ranked fourth and fifth. In 1915 the number of white women listing occupations in Buxton had declined from 11 percent in 1905 to 6 percent. By 1915 white women's occupations had shifted slightly from domestic or unskilled work to professional or semiskilled work. In 1915 white women's four major occupations were teacher,

Table 5.4. Occupations of Females, Buxton, 1915

Occupation	White	Black
Clerk	12	2
Farmer	12	—
Plasterer	—	1
Teacher	12	6
Bookkeeper	3	—
Postmaster	1	—
Dressmaker	1	4
Music teacher	—	1
Domestic	11	48
Agent	—	1
Telephone office worker	1	—
Cook	—	3
Wash woman	3	1
Waitress	3	1
Housekeeper	1	2
Assistant postmaster	1	—
Boarding house	1	1
Livery and hotel	1	—
Paper factory	1	—
Hairdresser	—	3
Field secretary, Wesley Convention	—	1
Packing house	1	—
Restaurant worker	—	1
Moving pictures	2	—
Midwife	1	—
Elocutionist	—	1
Hotel maid	—	1
Other	13	13
Total	81	91

Source: Iowa, Census of Iowa, 1915, Manuscript Population Schedules for Bluff Creek Township, Monroe County.

farmer, clerk, and domestic. It is presumed that most of the clerks worked in the company store. Next in importance were bookkeeper, laundress, and waitress.[32]

In comparing occupations of black and white women in 1905 and 1915, white women engaged in a wider variety of occupations than did black women and also occupied more professional and skilled positions. Former residents have stated that Buxton's schools were taught by both black and white teachers. The census data support those statements but also indicate that white teachers outnumbered black teachers by almost two to one. Overall, most women listing occupations fell into categories of traditional female work, such as dressmaking and domestic. Black women, however, more frequently fell into the unskilled or laborer category. In 1905, out of a total of

Employees standing outside the Buxton Savings Bank. From left to right are Olive Maddy, T. R. Cole, and Evelyn Cochran. Lottie Armstrong Baxter, daughter of Hobe Armstrong, also worked in the bank for many years. (*Courtesy of Wilma Stewart*)

123 black women employed, 70 were in unskilled occupations; in 1915, out of 91 black women employed, 61 listed unskilled occupations.[33] No doubt this situation stemmed from the fact that black women had less education than did white women and that there were fewer opportunities for black women to receive training in skilled or semiskilled positions.

The occupations of "boarding" and "boarding house" deserve special attention because the census data do not agree with information gained through interviews. In 1905 the state census included four white women and nine black women in Buxton who listed "boarding house" as their occupation. In 1915 two women, one white and one black, listed that occupation. In addition, the 1905 census included one white woman who listed her occupation as "boarding and sewing."[34] Given Buxton's population in 1905 of almost 5,000, it is not unreasonable that the community included thirteen boarding houses. It is not known, however, if these boarding houses differed in size and design from the company houses. It is possible that women who responded in 1905 and 1915 as operating boarding houses were doing so out of their homes. Dorothy Collier's maternal grandmother, Mrs. Reasby, essentially turned her company house into a boarding house. Widowed at the time, she kept boarders in the two bedrooms upstairs, while she utilized the living room for her bedroom. In some Iowa mining camps, women occasionally rented two adjacent houses from the company. Their family resided in one house and they filled the other house with seven or eight boarders. It is also known from interviews that many women, both black and white, kept boarders in their homes, but they apparently did not state this fact to the census taker. Because of this fact, the women who kept boarders, and who in turn performed work for wages within the home, are definitely undercounted in both the 1905 and 1915 censuses.[35]

The 1915 state census also includes information on income that provides additional data on Buxton's working women (Table 5.5). Although not all women listing occupations disclosed their income to

Table 5.5. Annual Income of Employed Females, Buxton, 1914

	White		Black	
Occupation	Income	Number	Income	Number
Clerk	$ 409*	11	$100	1
Farmer	1200*	4	—	—
Plasterer	—	—	300	1
Teacher	474*	9	437*	4
Bookkeeper	500	1	—	—
Postmaster	750	1	—	—
Dressmaker	300	1	250*	2
Music teacher	—	—	400	1
Domestic	168*	4	314*	32
Agent	—	—	250	1
Telephone office worker	450	1	—	—
Cook	—	—	325*	2
Laundress	200	1	172*	3
Waitress	383*	2	160	1
Housekeeper	—	—	250*	2
Assistant postmaster	750	1	—	—
Boarding house	650	1	550	1
Livery and hotel	400	1	—	—
Paper factory	300	1	—	—
Hairdresser	—	—	180*	3
Field secretary, Wesley Convention	—	—	720	1
Packing house	300	1	—	—
Restaurant worker	—	—	200	1
Moving pictures	250*	1	—	—
Elocutionist	—	—	450	1
Hotel maid	—	—	160	1

Source: Iowa, Census of Iowa, 1915, Manuscript Population Schedules for Bluff Creek Township, Monroe County.

*Denotes when income has been averaged.

the census takers, approximately half did. Women were asked how much money they had made the previous year. Although no distinction has been made in Table 5.5 between married and single females, the overwhelming majority of females who listed occupations, both black and white, were single. Taking all female incomes listed, the average income of employed females in Buxton in 1914 was $327.03. Reconstructed household data from the state census of 1915 indicate that most unmarried females in Buxton remained within the parental home until they married. Given that practice, and the fact that minor children, both male and female, were expected to turn their paychecks over to their parents, young women were bringing sizeable amounts of money into the parental home. Moreover, in some cases families included more than one working daughter.[36] (See Chapter 3 for a discussion of the wages of male workers in Buxton.)

Table 5.5 also shows the wage differential between black and white women. The average income of black women in Buxton who listed their 1914 income was $300.43, while the same figure for white women was $362.91. This means that white women averaged $62.48 more in 1914 than did black women. The differential may indicate some wage discrimination against black women. In many occupational categories, however, it is hard to compare precisely the wages of black and white women because not every occupational category includes both groups. Second, it is difficult to compare black and white incomes for occupations like domestic or laundress because it is not known if all women worked an equal length of time. These were occupations in which women worked by the day, and interview data suggest that this employment was often not full-time. Third, the occupations themselves indicate that, in general, black women held less-skilled positions than did white women. On the other hand, the category of teacher allows for a more precise comparison. According to income listed for teachers, black women averaged $437 in 1914, while white women averaged $474. This indicates that white teachers received $37 per year more than black teachers.[37]

Interview data also show that both black and white unmarried women worked outside the home, although they continued to live within the parental household. As a young woman, Agnes Erickson's Swedish-born parents sent her to Albia to learn the dressmaking trade. A short time later, in 1912, the company store needed someone to do alterations, so they hired Agnes and two other women. Agnes

A sixth-grade class in Buxton, 1907–1908. The teacher is Minnie B. London. (*Courtesy of Iowa Mines and Minerals Department, Des Moines*)

received eighteen dollars a month for that work. Although she had learned a skill, she commented that at that time, "I got started there and I didn't have hardly any school." Agnes also worked for a short time as a clerk, but she did not care for that position. She recalled that the store clerks worked long hours; the store opened at 8 A.M. and remained open until the miners' train arrived back in Buxton in the event that the miners wished to shop at the store before going home. Agnes next went to work at the company pay office, where she made out statements. Apparently she then received a substantial raise, as the state census of 1915 listed her yearly wage at $600.[38]

Soon after they finished eighth grade, both Nellie Lash King and Gertrude Stokes did laundry for Buxton families. Nellie King continued to help her mother at home, but she also did occasional washing for local residents. Sometimes she went to the John Baxter home to launder Baxter's shirts. Mrs. King explained: "If somebody wanted me to do wash or something, I would. People worked by the day, you know. You'd go to somebody's house, and work all day long." The pay was a dollar a day. When Gertrude Stokes finished eighth grade, she began doing laundry for railroad workers. "I would get their clothes," she said, "and carry them home and wash them and then I would carry them down—those men were engineers and brakemen—I'd leave them down there or sometimes I'd go down." Mrs. Stokes would often wait at the depot so that she could personally give the men their clothing. That enabled her to collect her wages at the same time. Otherwise, she noted, "I'd have to wait quite a while for my money."[39]

Some Buxton females went to work at an extremely early age. Odessa Booker remembered that at age ten she went to work for the Eric Brown family. Brown was a paymaster for Consolidation. "I used to go up there and baby sit and wash dishes," she recalled. "Got a dollar and a half a week." Later, when she was fifteen, Mrs. Booker went to work in the company store.[40]

Overall, the history of working women in Buxton parallels quite closely the history of working women in Iowa's other mining camps and in the Northeast. In most Iowa camps, women from all ethnic groups took in boarders as the major way of earning money. Many women, particularly the Italian-Americans, sold garden produce and dairy products and performed personal services, such as sewing, to supplement family income. Buxton offered a far wider range of jobs for women, however, than did most Iowa mining camps.

In comparing the work experience of Buxton women with that of women in other parts of the nation, the most precise comparison can be made with black women. Herbert Gutman, in *The Black Family in Slavery and Freedom, 1750–1925*, notes that in New York City in 1900 between 54 and 66 percent of black women listed occupations. The major occupations were service worker, washerwoman, and laundress. Elizabeth Pleck, in her study of black and Italian-American married women in New York between 1896 and 1911, notes that a higher percentage of black women have always worked outside the home. Pleck writes that in most American cities in 1900, the rate of employment for black married women "was anywhere from four to fifteen times higher than for immigrant women." Pleck asserts that "three dimensions of poverty—low family income, chronic male unemployment, and unskilled labor for men" led to the employment of black and Italian-American women. Between 1896 and 1911, most black women in New York City worked as laundresses and domestics. Many black women also took in boarders.[41]

The experience of black females in Buxton differed considerably from the experience of black females in the Northeast. It should first be noted that in Buxton few married black women worked outside the home, although many did take in boarders. Second, although unmarried black females had a higher rate of employment than did unmarried white females, that rate was far lower than rates for black females in other areas at a comparable time. Several things probably account for that fact. First, although many Buxton families were no doubt hard pressed to live on a coal miner's wage, that wage was higher than the wage of an unskilled worker in the industrial Northeast.[42] Interviews indicate that black families in Buxton perceived themselves to be making excellent wages. Also, there was little unemployment among black males in Buxton, since the men who were not able to find jobs there quickly moved on with their families to seek work in other coal camps. Second, Buxton provided fewer opportunities for women's employment, either married or single, than did a metropolitan area. No doubt every black married woman in Buxton had an opportunity to take in boarders, but the number of jobs available outside the home was simply less than what was available in a major city.

In determining family income and work patterns in Buxton, it is also necessary to examine the work habits of male offspring. In a sizeable number of Buxton homes, both white and black adolescent and young adult males worked outside the home and turned their wages over to their parents. At times, the number of working children per family was substantial. Robert Wheels remembered that in his family six of the seven sons worked as coal miners, although not all worked at the same time. Their father also worked as a miner. Wheels recalled that in his family all unmarried sons, before they reached age twenty-one, turned their paychecks over to their parents. The parents then gave each son a small amount for spending money. Wheels recalled: "Couldn't get nothing until you were twenty-one years old. When you come twenty-one, well you come to be a man. . . . All I did was work, and it was either my mother or father or next oldest brother, you know, . . . would pick up my check and things." The same practice prevailed in the Lewis family. All four of Lewis's brothers turned their paychecks over to their parents. Three brothers were miners and one was a mule driver.[43]

Buxton residents line up in front of Consolidation's pay office to collect wages. Note the superintendent's home on the right. (*Courtesy of Iowa State Historical Department*)

Herman Brooks also remembered that young men in Buxton did not keep their own wages. In the Brooks family, the father collected all his sons' paychecks on payday. Herman Brooks stated that the day he was twenty-one his father gave him "his time," meaning his wages. Brooks also contributed to his family's income by paying for his father's tools and blasting powder. He explained: "I never did let my father pay for anything. I paid for everything. My father . . . had to have powder and tools, you know, to work with. . . . I paid for everything and my father's money was clear." After Brooks reached his twenty-first birthday, he and his father alternated cars. The first coal car they loaded was credited to his father and the second one was credited to Herman, and they alternated that way through the day.[44]

In regard to the work habits of Buxton adolescents, there was apparently no difference between the experiences of black and white families. In Bodnar, Simon, and Weber's study of ethnic groups in Pittsburgh at the beginning of the twentieth century, the authors noted that black children did not contribute as frequently to their family's income as did children from Italian and Polish families. Although the Buxton study does not include comparable data on Italian and Polish families, when wage-earning practices of black Buxton residents are compared with those of white Buxton residents (including both foreign-born and native-born families), all groups evidence the same general patterns. In other words, children from black and white families alike frequently became wage earners during their early adolescence (in some cases even before this), and as a general rule turned their wages over to their parents until they reached twenty-one.[45]

A major concern of historians and sociologists who study family history is family stability. This issue is often raised in the study of immigrant families or of families migrating within the United States. Did relocation have an adverse impact on families, weakening family solidarity or causing the actual disintegration of family units? Did parental responsibilities and family relationships remain the same after relocation? What in fact did parents perceive to be their major responsibilities in regard to their offspring? Throughout Buxton's existence, the family played a central role in the lives of its inhabitants. Census data show that overwhelmingly Buxton residents continued to live in family units. Interview data indicate that Buxton families performed the traditional functions and responsibilities common to most families. As discussed earlier, blacks who migrated from the South to Muchakinock (and later to Buxton) came in family groups. Whites followed the same pattern. Boarding houses did exist for single male workers, but only an extremely small proportion of males lived outside of traditional family units.

Black families who moved to Buxton seem to have experienced only minimum difficulty adjusting to a new region and a new occupation, and they experienced less difficulty than did blacks moving to other sections of the country, particularly the Northeast. Black families who settled in Pittsburgh, for example, encountered difficulty in finding housing and employment. Both problems caused considerable stress that had far-reaching effects on the family's ability to adjust and prosper.[46] Black families arrived in Buxton under considerably different circumstances. Many families had been recruited by the company, which meant that jobs and houses awaited them. After the first group of black recruits arrived, they often wrote to family members and friends back home encouraging them to migrate to Buxton too, and they had every opportunity to approach the mine foreman, or possibly even the general superintendent, to inquire about jobs for the newcomers.

Black families did have to adjust to a new occupation, but several interviewees mentioned that their parents had little difficulty adjusting to a new line of work. When asked if her father, who migrated from Virginia, experienced any problems adjusting to coal mining, Mattie Murray answered: "No, he didn't have no trouble at all. He just did what they wanted him to do." She added that at the time her parents moved to Muchakinock her father could not write but that she and her brothers and sisters later taught their father to write his name.[47]

White families in Buxton seem to have been required to make few if any adjustments. Many had previously lived in other coal mining communities, so the women and children had some experience with camp life and the male breadwinners were accustomed to mine work. In fact, many white males were the second, perhaps even the third, generation of their family to work in coal mines. Numerous white families lived in company housing for a short time and then purchased acreages or small farms close to Buxton. The men then combined the occupations of farming and mining. While this increased their work load, it also gave them far more economic security. When the mines shut down, these families had a second occupation to fall back on. Among white families, the foreign-born probably faced the greatest adjustments. With most ethnic groups, however, a network soon developed as other families from the same European country settled in close proximity to one another.[48] (The role of ethnic groups in Buxton is discussed in Chapter 6.)

Interviews with former Buxton residents indicate that both black and white families demonstrated strong kinship ties. The Erickson family provides an example of this family solidarity. In the 1890s, Charles and Josephine Erickson emigrated from Sweden to Muchakinock, where Charles went to work as a coal miner. The Ericksons subsequently had three children, Agnes, Alex, and Dena. Throughout the lifetime of the parents, all family members lived close together. Alex became a coal miner and worked with his father, Dena married a coal miner, and Agnes worked for Consolidation for almost twenty years. Thus all family members had a common employer. Alex and Agnes never married and both remained in the parental home. Following their parents' deaths, they maintained a common home. Dena married, but she and her husband always lived next door to the parental home. When Buxton closed down in the early 1920s, the parents, along with Agnes, Alex, Dena, and Dena's family, relocated to Pershing, in nearby Marion County, where the two families continued to live side by side.[49]

Irene Goodwin's family also provides an example of family solidarity among Buxton's white population. The first generation, John and Mary Chambers, arrived in Muchy in the 1890s and moved to Buxton in the early 1900s. For the next forty years, three generations of the family remained active in Iowa's coal mining industry. The Chambers had four daughters, all of whom married and remained in the same vicinity. Irene Goodwin, granddaughter of the Chambers, grew up in Buxton surrounded by two sets of grandparents and many aunts and uncles. She lived with her grandparents on several occasions during her adolescent years. After Irene's marriage in 1919 to a coal miner, she and her husband lived in Buxton, and they later moved to Haydock. Throughout these years, members of Irene's family interacted frequently with one another.[50]

Black informants also related experiences that indicate family stability and solidarity. Many remembered that in times of hardship family members came to the assistance of one another. After the death of Susie Robinson's first husband, several family members assisted her. She related: "My father did mostly, . . . if I needed anything at all, and my brother, they were both working in the mines. If they thought I needed anything, well, they'd come give it to me." Dorothy Collier also remembered family members helping one another. For several years, as mentioned earlier, Dorothy and her brother stayed with their maternal grandmother while their mother worked as a cook. Later they stayed with an aunt in Buxton while their mother moved from Buxton to Cedar Rapids.[51]

For some black residents, a strong sense of kinship carried over into their adult years. Bessie Lewis's family moved from Buxton because her older brother decided that their father should have an easier life, that he "had been in the mines too long." The brother found his father a job and a house in Des Moines and then proceeded to move the family there. Mrs. Lewis recalled: "[My brother] said 'That's alright, I'm gonna take Dad out of the mines. Dad's been in the mines long enough. Many of the people,' he said, 'been killed in the mines.

. . . Now, he's worked all these years and he's successful and hasn't even been hurt." ' Later, when Bessie Lewis moved from Des Moines to Minneapolis, her mother relocated with her.[52]

Most Buxton parents viewed the family as the institution through which they both satisfied their children's material needs and instilled in them the proper social, moral, and religious concepts. Parents believed, moreover, that children should be raised to accept responsibility and to work hard, both in and outside of the home. Many informants, both black and white, discussed their early years and the attitudes their parents held toward all aspects of child rearing. Numerous black informants discussed the matter of discipline in some detail. These informants indicated that their parents took their role as disciplinarians seriously and that they firmly believed in the philosophy of "spare the rod and spoil the child." Informant after informant related the strict discipline and close supervision they received as children. Odessa Booker's parents, for example, were extremely strict with their children. Sometimes her brother would forget to bring in the coal before he left home. "He'd run home and try to get it in before he'd get a whipping," she recalled. "And they used to take his clothes off, you know, and take that belt and whip him naked and we'd just cry." She continued, "I remember one time Daddy put a lot of welts on his back. They didn't whip no clothes, they whipped you."[53]

Oliver Burkett also recalled that parental discipline was strict. At mealtimes in the Burkett family, everyone was expected to come to the table at the right time. If anyone arrived late, they did not eat. "If we were late, it was too bad," he recalled. "We should have been there. They knew what time we were gonna eat and it was the same time every day." Lester Beaman stated that in his family his mother did most of the disciplining. Beaman remembered getting "knocked down" a lot. He recalled that if he was talking to someone, one of his parents would "come in the room [and] you'd find yourself getting up out of the corner. Wham! [They would] think nothing of it. They would strip you down and tear you up. I never will forget that." Beaman concluded that all his brothers and sisters were disciplined the same way. "[My mother] didn't care how big they was, either."[54]

While mothers apparently assumed the major responsibility as disciplinarians, parents often conferred on family matters. Mattie Murray's parents held many conferences on family situations ranging from finances to child care. She related: "My father worked in the mine and I seen him many a day, my mother and father, settin' at the table in the kitchen havin' a conference. They'd conference 'bout the bills and things." Odessa Booker assumed that her parents agreed to present a united front to the children on all matters, as she remembered that her parents never quarreled in front of the children: "I never heard my mother fuss and argue with Daddy, you know. When they did their arguing, . . . I guess we were in bed."[55]

Black parents in Buxton believed it was their responsibility to supervise their children closely so that they did not get into trouble in school or with the law. Mattie Murray remembered that her mother warned all her children: "You pick up anything, and they don't belong to you and let 'em come and tell me about it, I'm gonna kill you almost." Murray added: "We didn't pick up nothin' either." Many black informants stated that their parents always backed up the teachers in the public schools. If parents found that their children had misbehaved in school and had been punished by the teacher, the child was punished again once he or she arrived home.[56]

For black children in Buxton, discipline was often administered in a communal way. Oliver Burkett related that children were frequently disciplined by adults other than their parents. Burkett remembered that when he and his siblings were at the Beaman home Mrs. Beaman felt no reluctance in punishing the Burkett children right along with her own. "If we were over there playing and done wrong, Mrs. Beaman would whip us, send us home. And she told us to tell Mama what she whipped us for. If we told [our mother], we'd get another whipping." Burkett added that if the Beaman children were at his home his mother did the same. Dorothy Collier also remembered that as a small girl she didn't dare "sass" any older person because they were certain to tell her mother and she was then punished for unacceptable behavior. She asserted that in Buxton, "everybody looked after . . . children, you know."[57]

The strict discipline administered by black parents also involved teaching their children to respect others. Mattie Murray's mother insisted that all her children show proper respect for their elders and that they refer to all older men and women as "aunt" and "uncle." Her mother would say "I better not hear you call none of them people by their names. They're your aunts and uncles. They're old enough to be your aunt and uncle."[58]

Parental supervision seemed to be particularly strict for daughters. Many black women remembered that their parents had main-

The George and Alice Neal family, taken circa 1910 in Buxton. The Neals appear with their children Dorothy and Harry. (*Courtesy of Dorothy Collier*)

tained close supervision over them until they married. When Susie Robinson was a young teenager in Buxton, she was not allowed to go out without her brother: "I couldn't go nowhere without my brother or some of them with me. Every time I walked out that door, my brother was with me, and I better walk in that door with him, too." Mattie Murray remembered that as a young girl none of the children in her family were allowed to be out after eight o'clock in the evening. Gertrude Stokes also recalled that parental discipline was strong in Buxton. She remembered that movies were shown at the YMCA but, she commented, "I didn't get to go every night 'cause I lived out in the country. Of course, my daddy wouldn't let me go 'cause he'd bring me down, he'd come down and take me to the movies, 'cause I couldn't come down there by myself. He wouldn't stand for that."[59]

Some black parents forbade their children to go into certain parts of Buxton. Dorothy Collier recalled that there were certain parts of town and certain homes that her mother did not allow her to visit. There were also some children with whom Dorothy was forbidden to play. Marjorie Brown also recalled that she was warned by her parents to stay away from certain parts of town. Hazel Stapleton's parents were also strict about their children's activities. Her family lived on a farm outside of Buxton, and her parents allowed their children to go into Buxton proper only infrequently. Her mother did not let the children "mix with everybody," and her parents enforced a strict curfew. Hazel recalled that as they got older, if they were not home at a certain time their mother came looking for them. She remembered: "She didn't have much trouble with us, with me, coming home on time, because the next time she wouldn't let us go. We had to stay home."[60]

Several black women related the matter of discipline and close supervision to the subject of sex education. Apparently in most black homes this subject was taboo. Odessa Booker recalled that her mother never told her anything about pregnancy or the birthing process. She and the other children believed that a stork delivered babies. When Mrs. Booker's mother was ready to deliver, she sent the children next door to the neighbors. Once, when Odessa was fourteen, she told her mother that she was getting fat. Her mother responded that she had been eating too much. A short time later another baby arrived in the Booker household. Lara Wardelin recalled that young girls were never given any sex education by their parents. She stated that when she was a young girl in Buxton young people simply found out about sexual matters on their own. "If you asked your mother something," she recalled, "she'd say, 'You're too young. You don't know nothing about it. You just wait until you get old enough and then you'll learn.' " Mrs. Wardelin's mother gave her some information, but most of what she knew of the subject came from her older sisters. Mattie Murray also stated that parents did not believe it was their responsibility to educate their children about sexual or reproductive matters. One respondent related that when she married in her middle teens, "Well, I was just a kid. I didn't know. I didn't think. I didn't know that men and women had to live together. I told [my husband] 'You go home.' " Her mother then told her that she had married the man and that meant she had to live with him. She responded by telling her mother, " 'No ma'am, he's going home.' I ran him home. I didn't know when they married they had to live together."[61]

The concern that black parents exhibited for their children's proper upbringing and correct behavior went hand in hand with the parents' aspirations for their children's future. If that future involved working for Consolidation, then fathers were often involved, particularly in helping male offspring to find jobs. As discussed in Chapter 3, in Buxton, as in many other coal communities, not only did young men take up their father's occupation, but fathers often provided their sons with the training needed for that work by taking them on as helpers, often taking them into the mines as soon as it was legally permitted. In 1900 young males in Iowa could work underground when they reached age fourteen. Once underground, sons usually remained with their fathers until they reached age eighteen, learning the skills needed to become independent miners. At eighteen, most young men had the choice of working independently within the mine or continuing to work with their fathers, and informants indicated that they often remained with their fathers. Some explained that they did so for monetary reasons, while others stated that they believed their presence lightened their father's work load. If a father had several sons who wished to be miners, he often trained all of them. As each younger son reached the age where he could legally work underground, the older son usually moved on to work independently.[62]

In Buxton, if sons wished to work for the company but in some capacity other than as a miner, the father was also involved. Because he knew the company personnel, he often played an intermediary role, approaching the company foreman or superintendent on behalf of his son. The same situation often applied to daughters; if they wished to work in the company store, the father often made the initial contact with the company. While these conditions affected both black and white families, perhaps they were most meaningful for the black population. The black father, in providing for his family's material needs as well as for his children's occupation, was exercising considerable authority, and he emerged as a strong figure both inside and outside the home. The roles assumed by black men in Buxton thus contrasted sharply with what blacks sometimes experienced in other areas of the country. In northeastern urban centers, for example, black males were sometimes unable to find employment, with the result that their wives went to work. By contrast, in Buxton most black women stayed at home and the black male emerged as the primary breadwinner for the family. Not only did the black male have authority within the home, but he also had considerable stature in the outside world, first in working as an independent workman and second in promoting the economic interests of his offspring.

Many black interviewees stressed that their parents had high hopes that their children would grow up to lead meaningful lives. Odessa Booker's father had intended for her to attend high school in Oklahoma while living with a grandparent, but the father's premature death in the mines ended those plans. Harold Reasby's parents had talked to him constantly about their desire for him to become a preacher. At the same time, he was repeatedly admonished to "stay out of trouble, treat people nice, regardless of what color they was." His father counselled him often: "Get a religion and live right and try to treat everybody alike." Some parents emphasized proper social behavior as a means of helping their children to succeed as adults.[63] Dorothy Collier's parents stressed proper table manners:

> My brother would always have to pull a chair for Mama, and we had to use the correct silverware. And a blessing was said. . . . And we had to use a napkin. We all had a little ring that you put it in. And always a tablecloth. I could only sit in a certain way. Since I was a little girl, I couldn't sit with my knees crossed. . . . . And my father was that way through his entire life. He was very particular about eating at the table and sitting correctly. When you came in the door with a hat on, the first thing he would say, "Did you forget something?"[64]

Vaeletta Fields recalled that her mother, Minnie B. London, had high aspirations for her children's education and success. Vaeletta believed that there was a strong motivating force within her family for her and her brother to do well. She related: "[That was] the atmosphere I lived in. . . . It was expected of you." When asked if most black parents in Buxton had similar aspirations for their children, she replied: "Well I guess that's the way it is every place. Every home is not the same and doesn't have the same model. I just don't know that . . . but we just felt like it was expected of us and thought nothing of it, you know, made no bones of it." Mrs. Fields related that George Woodson offered to send her to business college in Des Moines but her mother did not want to be obligated to anyone. Moreover, Mrs. London believed that teaching "was the only thing that a Negro girl could ever do." Mrs. London taught school for many years in Buxton and helped send her two children, Vaeletta and Herbert, to the University of Iowa. Vaeletta remembered that it was still difficult to get by, even with her mother's assistance. She related: "The order of the day was [that] you lived with a family for room and board and did little chores, you know." She added that the young men tended furnaces and cleaned up and waited tables at fraternities and sororities. Mrs. Fields graduated from the university in 1917 with a degree in English. Her brother also graduated from the same institution with a medical degree.[65]

Marjorie Brown's parents and grandparents also placed great emphasis on education and personal development. She remembered that her parents stressed education and culture and enjoyed a high standard of living. Mrs. Brown's family enjoyed considerable physical comfort. She did not remember, for example, any time that the family did not have a carpet on the floor or a piano. Her mother insisted that the family enjoy certain niceties, such as having a freshly starched tablecloth on the dining room table every day. Following the death of her mother when she was ten years old (her father had died a year earlier), Mrs. Brown and her brother went to live with their maternal grandparents in Buxton. They also valued education and constantly encouraged their grandchildren to do better in school. Every evening as they gathered at the dinner table, the grandfather asked the two children: "Well, what did you learn today?" Mrs. Brown related:

> Now that didn't mean that you learned to wash dishes; but what did you learn, what did you see, what made it an interesting day? What did you read and what did you learn? And if you said nothing, if you hadn't learned anything, if you sat there dumb, then he would tell you how you were wasting your life. Life meant you should learn. And I thought all the people did that.[66]

Certainly the aspirations of black parents were also reflected in the fact that a number of black Buxton youths did go on to high school and college. Vaeletta and Herbert London attended high school in Oskaloosa before entering the University of Iowa. In 1903, Eva Bates graduated from Oskaloosa High School. Edward A. Carter graduated from Oskaloosa High School before 1900, but during the time his parents lived in Buxton he attended the University of Iowa. According to the *Bystander*, one of the most popular young black men in Buxton in the early period was Linford Willis. Willis took part in many local activities, serving as president of the Baptist Young People's Union, director of the Mt. Zion Baptist Church choir, and manager of the Buxton Cornet Band. In September 1902 the *Bystander* reported that Willis was about to leave for Des Moines to enter Highland Park College. In 1911, Willis returned to Buxton with a degree in dentistry and set up a private practice. Sometimes the accomplishments of black young people took a different form, however. In February 1902 the *Bystander* noted that Margaret M. Coleman, "one of Buxton's well known young ladies has recently been granted a copyright for a book, of which she is the author. The title is 'The Past, Present and Future of the Negro Race.' " The article noted that Miss Coleman had also composed several songs.[67]

Regardless of race and ethnicity, however, coal mining families shared a common concern: the safety of the men in the mines. Coal mining had long been recognized as one of the most dangerous occupations, not only in regard to injury and death but also to long-term disabilities resulting from working conditions. Notices such as the following frequently appeared in local newspapers:

> Saturday was an unfortunate day at Mine No. 10. A young Swedish man, J. C. Carlson, was caught with falling slate and both legs and arms were broken. He also received internal injuries. He died Sunday morning and was buried by the union Monday. Mr. Carlson was a very nice young man and had only been in America for one year. His parents are in Sweden.[68]

Odessa Booker recalled vividly the day her family was notified of her father's death in Buxton. She related that on that particular day the men were rushing to do their work:

> Just before noon a slate fell on him and hit him right back, right back there, and broke his neck. . . . I can remember that, you know, 'cause they all—in the mines you could tell when anyone got killed or hurt in the mines. . . . The whistle would blow, you know, they'd sound off and you'd say "Oh my goodness, somebody got hurt," and we didn't know it was Dad. Mama was washing that day, . . . we were washing clothes, and finally Dr. Carter . . . he

came [and] told my mother, 'cause some of them didn't want to come, so he came and told her about Father getting killed in the mines.[69]

Gertrude Stokes remembered that when she lived in Buxton the train engineer announced a mine fatality. She recalled, "You know, when somebody got killed in the mines, Ed Kaizer [the train engineer], he had a certain whistle he used and the women would be at home, but they'd get busy then. They'd . . . go down and meet that train and see who got killed."[70]

For many Buxton women, the dangerous conditions in the mines were something that they simply had to take in stride. Some informants stated that they didn't worry much about the danger because it was just a fact of life for coal mining families. Other women responded that the probability of their husbands and sons being hurt or killed in mine accidents was a thought that never left them. For many women, like Gertrude Stokes, the sound of the train whistle or the mine whistle was an agonizing experience.

For individuals who lived in Buxton, life was determined to a great extent by responsibilities, expectations, and financial conditions imposed by the immediate family circle. While the occupation of coal mining certainly created some tension and stress within families, particularly because of its dangerous nature, there is no evidence that it jeopardized the solidarity or cohesiveness of family units. For the most part, family relationships seemed solid and enduring among both blacks and whites. When hardships and tragedy struck, families seemed to cope with the difficulties, often with assistance of other family members. Because of the social and economic opportunities available there, most black families adjusted, and they accommodated their lives to that environment with considerable ease. These transitions stood in contrast to the adjustments and difficulties experienced by black families in other parts of the country. For white families, many of whom had already experienced life in a coal camp, the adjustments were even easier, as life in Buxton closely approximated what they had known elsewhere.

For married women in Buxton, perhaps life changed least of all. While Buxton provided families with good housing and steady wages, women's work within the home remained much the same. For black women with large families, household tasks seemed endless. Married women found their days filled with the domestic duties of child care, meal preparation, housecleaning, laundering, mending, and sewing. Many women, moreover, took in boarders. This added considerably to their domestic work load, but it made the women's monetary contribution to the family substantial.

Children were also important wage earners in most Buxton homes. Children typically went to work when they finished the eighth grade or when they reached age fourteen, and many had part-time employment before age fourteen. Children routinely turned their wages over to their parents, so if a household contained several teenage children or offspring in their twenties it might have four or five individual incomes. Children also performed many tasks within the household. In this sense, most family members served dual roles: they accepted work responsibilities within the home and worked as wage earners outside it.

6

ETHNICITY

IN 1902 the editor of the *Bystander* observed: "Buxton is a new mining town. . . . Old Muchy used to be the most thoroughly colored town in Iowa, but now Buxton wears that mantle, for she is destined to be when completed, the largest, nicest, and most modern mining town in Iowa and perhaps in the United States." In the next eight years, the *Bystander* editor was to see his prophecies come true. In 1905 he wrote: "Buxton is the colored man's mecca of Iowa;" and in 1910, he referred to "the far famed Buxton, which has been truly styled the Negro Athens of the Northwest."[1] With perhaps less-colorful rhetoric, this same view of Buxton was expressed by almost all black former residents interviewed for this study. Blacks reported that Buxton was a special experience for them, virtually a racial utopia and one that they never experienced again. But while Buxton was biracial, it was also multiethnic. People from many foreign countries lived and worked side by side in Buxton. This chapter will explore the experiences of these ethnic groups, first by examining the ethnic communities and second by exploring the interaction between ethnic groups in the areas of work, housing, education, and public accommodations. Throughout the chapter, the experience of black residents will be emphasized, partly because of their dominant position, but also in an effort to answer the question: did racial or ethnic discrimination exist in Buxton?

Although Buxton remained a biracial, multiethnic community throughout its existence, the ratio between groups underwent considerable change. In 1905 black residents accounted for 55 percent of Buxton's total population of 4,921. By 1915 blacks still comprised the largest single ethnic group, but with 40.4 percent of the total population of 4,518, they no longer represented a majority. On the other hand, the foreign-born never accounted for more than 14 percent of Buxton's total population, and in 1905 it constituted only 7 percent. This underestimates the importance of Buxton's immigrant families, however, since American-born children of immigrants were not included in these figures, though they were in fact members of ethnic communities. By 1915 the first and second generations of European immigrants accounted for nearly one-third of Buxton's population.[2]

Buxton's three largest and most significant European ethnic groups were the Swedes, Slovaks, and immigrants from the British Isles (Table 6.1).[3] By 1915 the number of Italians had increased considerably, but because there were only a few Italians in Buxton in 1905 they have not been included in this analysis. Similarly, Buxton included immigrants from Belgium, Bohemia, France, Germany, Norway, and Russia, though these groups were small both in 1905 and in 1915.[4] It should be noted that the majority of Buxton residents were native-born and had native-born parentage. Most of Buxton's residents, moreover, were native-born Iowans. In 1905, 42 percent had been born in Iowa, and in 1915, 54.6 percent.

Of special interest are the birthplaces of Buxton's largest ethnic group, the blacks (Table 6.2). Black Americans were the largest ethnic group in Buxton and constituted the majority of its population until 1910. No doubt because of this, blacks had the most well-developed

Table 6.1. Percentage of Swedish, Slovakian, and British Isles Immigrants in Total Population, Buxton, 1905 and 1915

	Percentage of Total Population	
Birthplace	1905	1915
Slovakia	2.0	4.8
Sweden	3.0	1.9
British Isles*	2.0	2.2

Source: Iowa, Census of Iowa, 1905 and 1915, Manuscript Population Schedules for Bluff Creek Township, Monroe County.

* Although the British Isles do not represent a single ethnic group, "British Isles" is used as an ethnic category for our purposes here. It is composed of people from England, Scotland, Wales, and Ireland. The people share a language, one that is different from that of the Swedes and Slovaks, and one that was spoken in the United States.

Table 6.2. Birthplace of Black Residents, Buxton, 1905 and 1915

	Percentage of Black Residents	
State of Birth	1905	1915
Virginia	33.0	27.0
Iowa	24.0	41.0
Missouri	8.9	8.4
Alabama	6.6	3.7
Tennessee	4.8	2.9

Source: Iowa, Census of Iowa, 1905 and 1915, Manuscript Population Schedules for Bluff Creek Township, Monroe County.

ethnic community there. At the center of that community was the black church. Buxton contained at least eight black churches: three African Methodist Episcopal churches, three Baptist churches, a Congregational church, and a Church of God. The *Bystander* reported on February 6, 1914, that 538 people had attended the Mt. Zion Baptist Church and 375 had attended St. John's AME Church that week.[5]

The black churches played important spiritual and social roles in the lives of Buxton's black residents. According to interviews, most families belonged to a church and attended regularly. Hucey Hart remembered that his mother "was quite a churchgoer" and that she spent a good deal of time there. Hart's family attended the Mt. Zion Baptist Church, and Hart recalled that Wednesday night they attended a prayer service that would last three or four hours. On Sunday the families took food for a potluck dinner and remained at church all day. The minister preached in the morning and again in the afternoon. The Baptists also held church services on Sunday evenings.[6] Elmer Buford recalled a similar experience: "I went to Sunday school, then my grandmother'd catch me and take me back into church. Yeah, yes. Never missed a Sunday." Odessa Booker explained: "I remember at night we'd go to church. . . . We had to take a lantern, you know. . . . We'd carry a lantern so we could see our way home."[7]

Music was an important part of the black churches' activities. Both St. John's and Mt. Zion had large choirs, and in April 1902 St. John's organized a nine-member black orchestra to accompany its choir. Marjorie Brown's mother, Melvina Lee, played the organ at St. John's and directed the choir. Mrs. Lee had graduated from Leon High School and was described by her daughter as an accomplished

A summer gathering at St. John's African Methodist Episcopal Church. The long, rooflike structure on the right is a brush arbor, built to provide shade for outdoor events. (*Courtesy of Iowa State Historical Department*)

musician. Linford Willis directed the Mt. Zion choir, which had twenty members. Many former residents remembered that the black churches had excellent choirs; as a young man, Carl Kietzman would enter the black churches by the back door to listen to the music.[8]

Both St. John's and Mt. Zion had a wide variety of organizations for their members. As mentioned earlier, in March 1902 the young people at Mt. Zion organized the Baptist Young People's Union. Hucey Hart belonged to this group. The BYPU met before the Sunday evening church service. In September 1902 the *Bystander* reported that "One of the leading organizations and auxiliary clubs of Mt. Zion Baptist Church is known as the 'Gold Bugs.' The club is rendering valuable services to the church, both spiritually and financially. The club is composed of earnest Christian workers." The article concluded that the women of the organization had presented a play, "Ye Old Folks and Children's Concert," and in the near future they hoped to present a new play, "Two Aunt Emilys." A month later, Mt. Zion's young people organized a literary society, the Paul Dunbar Society, which met every Monday night. In January 1911 the young men at St. John's organized a Young Men's Forum where, the *Bystander* noted, "young men can meet on Sunday afternoons and discuss different topics and spend a pleasant hour socially." Both churches had Sunday schools that held public programs at various times and also presented special programs at Christmas.[9]

Sometimes the black churches entered the realm of secular education. In February 1902 St. John's presented a program on "Child Training" in which three teachers from the Buxton public schools—Misses Trueman, Owens, and Long—took part. The *Bystander* observed: "The subject was well handled and many facts were brought up. Mothers and trainers of children who attended this service will surely profit by many of the things said." In 1910, the Mt. Zion Baptist Church set up a night school in which Professor C. W. Rodgers conducted a school for approximately forty people. At the end of the session, the participants put on a program of songs, declamations, and a debate.[10]

Voluntary associations comprised another element of the black community in Buxton. Blacks there belonged to at least forty distinct lodges, clubs, musical groups, political associations, and sporting teams. These groups produced a rich and varied social life for Buxton's black residents. The men's black fraternal lodges met in the main YMCA. Some lodges that existed in Muchakinock had transferred to Buxton. In October 1902, when Odd Fellows Lodge Number 220 was to be relocated in the new community, the *Bystander* noted that "A fine program will be rendered and a bounteous banquet supper will be spread." The Odd Fellows lodge seemed to be the most popular lodge, but black men also belonged to the Masonic Order, the Benevolent and Protective Order of Elks, and the Knights of Pythias. In 1902 the Buxton chapter of the K.P. had fifty members.[11] Herman Brooks, who belonged to the Masonic lodge, recalled perhaps the best part of lodge meetings: "At the meetings we used to have a big time, you know, and after the meetings we would serve drinks and sandwiches, and one thing and another." The lodges had annual days and, Brooks said, "They turned out. They had annual days when they had a band, the Buxton Band, and they'd parade and go to church and have a service."[12] Dorothy Collier also remembered lodge turnouts:

> The men's lodge would turn out. . . . There would be regalia. Oh, I tell you, the hats, big white hats, with big red plumes that hung down, and they had

the coats. . . . The horses had the big things that hung down the side and, oh, how they pranced. . . . The big sword all decorated, and the band leader, he would have the big gold thing that went around . . . . 'Course I was afraid to get close to them. I would peek over the hill and watch them, you know.[13]

These turnouts also occurred at other times, such as for the funeral of a lodge brother.

Black men in Buxton took part in a host of musical societies, choral groups, choirs, and bands. These included the YMCA Men's Glee Club, Professor Jackson's Braves, Dick Oliver's Band, and the Buxton Cornet Band, which was at one time a thirty-six–piece group and the best-known musical group in Buxton. In August 1911 the *Bystander* reported that the band "gave their first concert [of the year] last Saturday evening. A large crowd stood around on the walks and in people's yards listening to the sweet strains of music." The Buxton Band also took to the road. According to the *Bystander*, "The band is well known throughout the State of Iowa and the West. It was the only colored band at Omaha and Council Bluffs to meet the 51st Iowa Regiment from the Philippines over two years ago." Band members belonged to the National Musical Union of America.[14]

Although both black and white males took part in sports, black males appear to have organized the greatest number of athletic groups. Participation sports included baseball, football, basketball, tennis, swimming, handball, bowling, roller skating, and horse racing. In addition, the Gold Palace Gun Club was organized in 1903 with seventeen members. Herman Brooks recalled the Buxton basketball team playing teams from other camps: "Practically all of them came to Buxton to play . . . because we had the biggest gym, . . . but we never played a colored team." Many of the sporting events were organized and held at the main YMCA in Buxton, but bowling and roller skating were done elsewhere.[15] Bessie Lewis recalled that the racetrack was located near the ballpark. Hobe Armstrong, with his Thoroughbreds on nearby farms, was apparently the major figure behind horse racing in Buxton. The Langlois sisters managed the roller skating rink. Elmer Buford remembered some boxing in Buxton, and one fighter in particular by the name of Pep Webster. A physical culture club met regularly at the main YMCA for which "Y" officials charged fifty cents.[16]

Former Buxton residents agreed that the most popular spectator sport was watching the Buxton Wonders play baseball. Admission was fifty cents and local officials handled the concessions. The Buxton

Cornet Band regularly played at the games and occasionally the crowd followed the band to the town park for a concert after the game. As the *Bystander* observed in June 1902, "Last Sunday afternoon a game of ball was played between Lockman [a neighboring coal camp] and Buxton. . . . Immediately following the game the band struck up in [the] park, and the gay throng retired to the park where they enjoyed themselves eating ice cream and listening to the music."[17]

There were two distinct Buxton Wonders teams. B. F. Cooper organized the first Buxton Wonders, and this team played at the Amphitheatre, a ballpark in south Buxton. Mike Onder recalled watching the Wonders from special seats: "They had a great big ballpark, all fenced off. You'd pay to get in there, but then there was some trees right along Simm's farm and the kids—each kid had a tree when they played . . . and then you could look right over that fence." Earl Smith recalled that the first Buxton Wonders broke up in 1914 and the ballpark was torn down. Ed Peterson, a white Buxton resident, later reorganized the Wonders, and the second team played at a ballpark in northwest Buxton not far from downtown and near the stockyards. Later, in the 1920s, the team was known as the White Sox.[18] Herman Brooks pitched and played centerfield for the Buxton Wonders. He remembered both the Wonders and a younger YMCA team, and he claimed that the YMCA team was better than the Wonders at one time. He related, however, that B. F. Cooper "found out that the younger players were the best, . . . so he began to take one or two from the YMCA, put them on the Buxton Wonders."[19]

Some former residents recalled that the Consolidation Coal Company helped support the Buxton Wonders by donating land and building bleachers for the ballpark and also by buying uniforms. Others recalled that the managers, not the company, supported the Wonders. For the most part, players paid for their own uniforms, although they sometimes received money from the admission receipts. Most team members were miners and generally played ball only on weekends. If members took an occasional road trip that caused them to miss work, the company apparently overlooked this fact, but it is not clear whether the company paid wages to team members on their days off. Former residents recalled that teams from Chicago, Kansas City, and even Birmingham, Alabama, played the Wonders in Buxton.[20] The Buxton Wonders always had both black and white players, and their reception in nearby towns was not always pleasant, as the following newspaper comment makes clear:

The Buxton Wonders baseball team. George Neal is in the second row, far right. (*Courtesy of Dorothy Collier*)

The Buxton ball team wrote the Albia ball team (white) asking them for a game . . . and what do you think? A reply came something like this: "Albia, Iowa—Your communication received, and will say; we will not play against a colored team, any time you can bring us a white team [we] will play you."[21]

Along with sports and social organizations, black men also participated enthusiastically in politics. Political activity focused on race issues and on the Republican party. In November 1902 the *Bystander* reported: "Tuesday is election day and all the qualified voters in Buxton will vote the straight Republican ticket. They talk that way and walk that way." Apparently the Republicans were successful, for the *Bystander* noted that "the Republicans of the county snowed the Democrats under in this election. This county heretofore has been a Democrat county, but since Buxton has sprung up the whole politics of the county has changed." The same newspapers reported on April 10, 1908, that "the Buxton Republican Club was organized last Friday night."[22] Reports in the *Bystander* indicate that both Republicans and Democrats held rallies in Buxton. In November 1910 the *Bystander* reported that Governor B. F. Carroll had recently spoken at a Republican meeting in Buxton, that N. E. Kendall (an Albia attorney and state legislator) was scheduled to speak at another Republican meeting, and that the Democrats were planning to have a rally in the YMCA auditorium.[23] Republican politics in Buxton sometimes reached out into the state and the nation too. In April 1915 the *Bystander* reported:

To Buxton, Iowa, a colored mining camp, belongs the honor of organizing the first club (Cummins Club). Now, therefore be it resolved that we hereby permanently organize the "Cummins for Presidency Republican Club" No. 1. And that we hereby pledge to Hon. Albert B. Cummins, U.S. Senator from Iowa, our support for his nomination as president of these United States by the Republican National Convention of 1916.[24]

Buxton's blacks sometimes brought racial issues to the attention of lawmakers in the state capitol. In 1909, George Woodson noticed that a bill introduced into the legislature would prohibit black lodges in the state from taking the same names and regalia as white lodges. Woodson sounded the alarm and the bill was defeated in committee.[25] Three years later Woodson ran unsuccessfully for the office of state representative from Monroe County.

Black women in Buxton also organized social clubs, literary organizations, and fraternal lodges. In 1903 the *Bystander*'s Buxton correspondent noted that "the Buxton ladies are progressing. . . . The latest organization is the Ladies' Industrial Club. They have regular weekly meetings, each lady bringing some kind of work, they exchange ideas, assist each other and so on—then they lunch!" In January 1906, Buxton women formed the True Reformers. Described as a secret society, the group provided entertainment for Buxton residents during the Christmas season. Early in Buxton's history, black women also organized a chapter of the Federated Black Women's Clubs of America. In April 1907 it joined with the Ladies' Industrial Club in passing resolutions approving the efforts of the Buxton Ministerial Alliance to close the saloons on the outskirts of town. In 1907 the *Bystander* also reported the activities of the Sweet Magnolia Club. Marian Carter, daughter of Dr. E. A. Carter, recalled that her mother, Rosa, belonged to the Silver Leaf Club, which was regarded as one of the most elite black women's groups in Buxton. During 1909, 1910, and 1911 the *Bystander* also reported on the organization and activities of Phi Delta Phi, the Fannie Barrier Williams Club, the Fidelity Club, the Mutual Benefit Literary Society (for both men and women),

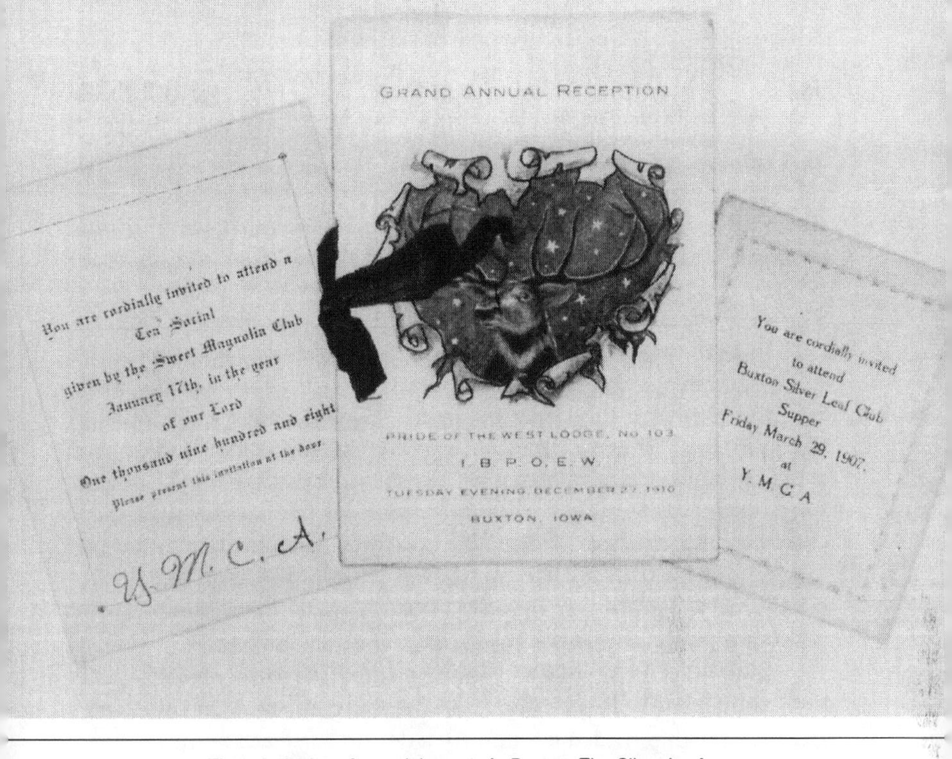

Three invitations for social events in Buxton. The Silver Leaf Club and the Sweet Magnolia Club were prominent women's groups. (*Courtesy of Iowa Mines and Minerals Department, Des Moines*)

the Etude Music Club, the Self-Culture Club, and the Progressive Women's Club, all organized and supported by black women.[26]

Black women also belonged to at least five lodges: the Eastern Star, the Household of Ruth, the Court of Calanthia, the Knights and Daughters of the Tabernacle, and the Virginia Queen's Court. Apparently both the black lodges and the social clubs maintained close relations with the black churches in Buxton; before the main YMCA was completed, the black lodges held their meetings in the black churches. Every year the women's lodges, like the men's, held a turnout day, and the black social clubs apparently did the same. On May 10, 1907, the Sweet Magnolia Club attended the First Congregational Church, where Rev. A. L. DeMond preached their annual sermon.[27]

While many black women's organizations were primarily social in nature, their members also addressed serious contemporary political and educational issues. Topical issues of the day, such as the proper role of modern women and the prospects for the black race, were discussed in the women's clubs and in Buxton's literary societies. Some of these issues even found their way into print from the hand of Buxton residents, as was the case with Margaret M. Coleman's book, *The Past, Present and Future of the Negro Race*, mentioned earlier.[28]

In 1916, when women throughout Iowa were struggling to secure women's suffrage, Buxton's black population indicated their support. In June 1916 the *Bystander* observed: "Saturday was a big suffrage day in Buxton. Women, children, and men came together and formed a big suffrage parade, headed by the Buxton concert band. After a long, enthusiastic march, they proceeded to the auditorium where they were entertained by two able speakers." Unfortunately, the *Bystander* did not include the number of people who took part in the parade. It is significant, however, that Buxton's black women were making a public statement about their views on suffrage.[29]

Sometimes black women combined their scholarly interests with their church work. In January 1910 the night school at the Mt. Zion Baptist Church provided a forum for a debate between two black men and two black women. The subject was: "Resolved that women should be educated equally with men." The *Bystander* reported that "This was a hard fought battle. The negative struggled hard to convince the judges that woman was inferior to man, and should not be educated equally with him." The judges, however, awarded the debate to the women. Sometimes black women made presentations on their own. In November 1911 the Mission Circle at the Mt. Zion Baptist Church met at the home of Mrs. B. G. Woodward, and the *Bystander* reported that "a very good program was rendered, the cream of which was a paper read by Mrs. C. R. Foster, subject 'Do Men Respect Women as They Should?' Many of the sisters spoke on the subject!!!"[30]

While many black Buxton women took part in clubs, lodges, and church organizations, not all did so. Family size undoubtedly determined to a large extent the amount of time that married women spent outside the home. Odessa Booker and Mattie Murray's mothers, for example, did little but care for their large families and perform the many domestic tasks made necessary by those families. No doubt family income also affected the extent to which black women participated in activities outside the home. Presumably the higher the family income, the more frequently women were able to hire help for both child care and housework. For example, Rosa Carter, whose husband was a company physician, participated in many social activities outside the home. According to the 1915 state census the Carters' income for 1914 was $3,000. By comparison, some black households earned only around $300 to $400 in 1914.[31]

The social life of Buxton's black residents paralleled the activities of midwesterners in general. In his study of small-town life in the Middle West, *Main Street on the Middle Border*, Lewis Atherton writes that small-town residents began "to participate in a national trend toward organizational activities in the late nineteenth century." Atherton believes that this behavior became most pronounced in the early twentieth century; in fact, he labels it the "twentieth century cult of joining." Atherton explains that this development took place because people were increasingly unable to identify with the community at large. As towns grew larger and populations became more mobile, people joined organizations to give themselves a feeling of still belonging. Where people had previously identified with the total community, after 1900 they faced an increasing fragmentation of social life and reacted by joining countless organizations. Women began to join the Federated Women's Clubs of America, garden clubs, child study clubs, and literary societies. The Women's Christian Temperance Union organized in the 1870s and continued to be extremely popular after 1900. Men joined church groups, service clubs, and booster clubs. Americans became a nation of joiners, and Buxton residents were indeed a part of this organizational life.[32]

It is instructive to note that organizations in Buxton were not the type that black people had known in the South, where a black organization was automatically accorded second-class status. Black clubs, black churches, and the black YMCA in Buxton had not stemmed from local racial discrimination, nor had they resulted from the "separate but equal" doctrine handed down by the United States Supreme Court in the late nineteenth century. According to black former residents, black churches existed in Buxton because black residents wanted them, not because the blacks had been shut out of white churches.

While Buxton was the home of many gender-related organizations, the black community as a whole also sponsored community affairs. Special events in the black community included national holidays, such as the Fourth of July and Labor Day. In 1907 the *Bystander* proclaimed that:

> July 4th was a day long to be remembered in Buxton. Work was laid aside. . . . Long before the appointed hour, 1:30 p.m. the [parade] line was formed in front of the business organizations of Buxton. The YMCA and Sabbath Schools were represented in the parade [and all marched] through the streets of Buxton thence to the park. . . . Judge M. A. Roberts . . . was the speaker of the day. . . . Miss Daisy Lee read the Declaration of Independence. . . . At 9:00 p.m. a brilliant display of fire works was had.[33]

The miners marched in traditional parades on Labor Day, and blacks in Buxton celebrated Emancipation Day:

> On Saturday, September 21, Emancipation Day will be celebrated in Buxton. . . . All business will be suspended so that all can enjoy the celebration. At 11 o'clock there will be a grand parade, headed by grand marshall and assistant, a platoon of police and the Buxton Cornet Band. There will be speaking. . . . [and] music by the A.M.E. choir and the Cornet Band. There will be a roasted calf and a pig, there will also be sports of every kind. . . . All of Des Moines is expected.[34]

Former residents also remembered community events in Buxton. Susie Robinson recalled that Buxton had a "fine big park" and that Saturday night was the "young folks" night. She related that a merry-go-round ran there every Saturday night and that it was just like a carnival.[35]

In addition to organized activities, black families continually hosted private parties and events, which were well reported in the *Bystander*. In January 1907, for example, the *Bystander* included an account of a card party hosted by Mrs. Resse Trigg. It noted that after playing whist and awarding the prizes, guests enjoyed a "dainty two course luncheon . . . which consisted of peanuts, sandwiches, pressed chicken, banana salad, and pease in crullers, ice cream and assorted cakes." The guests spent the remainder of the evening dancing.[36]

While many men, women, and children took part in these organized social activities, the day-to-day activities of black family life included a wide range of informal or unorganized activities. Many former residents remembered that in the evening family members played cards, checkers, dominoes, and other games. They remembered their childhood days fondly. Hazel Stapleton recalled: "Yes, all of my days there, we just had happy days there, 'cause Mama never allowed us to spat with each other. We just had lots of fun. In the summertime, every Sunday we would go hunting flowers. The boys would go goose berry huntin', and with fall . . . we'd go . . . to gather walnuts. Just had lots of fun." Stapleton recalled that because they lived on a farm she and her family spent much time together. Outside games played by Hazel and her sisters included Annie Over and Drop the Handkerchief. Hazel's mother taught her to play the organ, and as a child she had dolls to play with, and her brothers had wagons, skates, and sleds.[37]

Former black residents also remembered many different outside activities in Buxton. Hucey Hart related that Buxton had a big reservoir where young people went swimming in the summer and ice skating in the winter. As a youngster, Hart had great enthusiasm for hunting:

> Oh yeah, hunting was good. See, why hunting was so good down there, they cut these trees down and piled up the brush, and, see, rabbits stayed under the brush pile. [Someone would] get down and shake the brush and about four or five [rabbits were] liable to come from under there. They didn't have no law. You could hunt any time you wanted.[38]

Paul Jackson recalled that after school "we'd go down to the woods and shoot slingshots and cut our initials in a tree and pick up Indian arrowheads." As a child in Buxton, Odessa Booker and her siblings went roller skating, ice skating, and sledding. "We made our sleds," she recalled, "what sleds we had, you know. We'd make kites, you know. We did a lot of that. Yeah, kids played so much different then, and you found a lot of things to do. You made your own kites and you played jacks and . . . we'd even make wagons."[39]

Former black residents also remembered that their parents enjoyed informal get-togethers. Hucey Hart related: "The parents, they usually got together, say about five or six of them, they'd be sitting somewhere around the shade tree, drinking their beer and home brew and stuff like that and talking about the mines, about 'I done this and that' and 'You know that old [room] they give me down there, man, there's nothing but a whole lot of rocks,' and stuff like that." Lester Beaman's parents "would have people [in] just like Saturday night. . . . Maybe there'd be ten families come to our house on Saturday night. They'd have this home brew and they'd have food and stuff like that." Odessa Booker's parents did not socialize much with other families, and her mother seldom went anywhere because of her work at home. Odessa explained that "She stayed home and fed the kids, you know, so we didn't do a whole lot of going out . . . and mixing around." Odessa did remember, however, one standard social practice that most families engaged in: inviting the preacher and his family to

Sunday dinner. She commented: "You know, in those days, the preachers used to come and help themselves. On Sundays they'd have to fix dinner for the preachers and they'd come eat."[40]

The Swedes in Buxton formed the most physically separate community. In contrast to the other ethnic groups, which were somewhat scattered throughout the town, the Swedes resided in East Swede Town and West Swede Town. Most Swedes owned their own homes rather than renting company houses, and they paid fifty cents a month to rent a lot from Consolidation. The Swedes were also the most numerous nonblack ethnic group, as Jeanette Adams remembered: "Oh, yes, there were a lot of Swedes. Yes, they had their own church and everything. Yes, it was quite their own town."[41]

Swedish churches constituted the center of the Swedish communities. Wilma Larson Stewart remembered that the church was the most important institution among Buxton's Swedes and that families attended church together. Ministers conducted services in both Swedish and English. Agnes Erickson remembered that the Swedish Methodist Church was located in West Swede Town and that Ebenezer Lutheran, the larger of the two, was located in East Swede Town.[42] The Ebenezer church had been located in Muchakinock before being moved to Buxton in 1902. A history of the Iowa Augustana Lutheran Synod included the following account:

> An unprecedented case exists, whereby a whole congregation moved from one locality to another. It has actually happened in that the congregation in Muchakinock has disappeared but it is again found in Buxton to where the church building has also been moved.[43]

In the process of moving the church, the congregation remodeled and enlarged it, and later they added a room to be used for meetings of societies and festivals. Ebenezer contained a Sunday school, and church records indicate that an average of forty-three children attended each year. The congregation also established a Ladies Aid shortly after moving to Buxton. Membership in the church averaged eighty-nine, and, as would be expected, most members were coal miners, although three were farmers and three were businessmen. Ebenezer did not have a resident pastor during the time it existed in Buxton but generally shared pastors with Centerville, Hiteman, and Albia.[44]

Ebenezer church officials also held a Congregational School each summer (except 1911) for children in the congregation. The school ran from 9:00 A.M. to 3:00 P.M. and lasted for ten weeks. Agnes and Alex Erickson remembered their experience in the school:

> They'd bring a student from Rock Island. Yeah, Augustana [College] . . . and he preached on Sunday, and he taught Swede school through the week. But we didn't have arithmetic or geometry in Swede [school] because we got that in American school. But we got grammar and reading and writing [in Swedish] and stuff like that.[45]

The Swedish summer school also included religious instruction.

The Swedish Lutheran church also sponsored groups and events for their young people, including Luther League. In the summer, the church held ice cream socials. Wilma Larson Stewart remembered that "Many young people from, not only East Swede Town, but from West Swede Town would come." These socials were held at the church. Adolph Larson recalled socials on Saturday nights: "That's where the young folk would come to social[ize]. We'd dance afterward out in the yard."[46]

Confirmation in the Lutheran Church was a significant rite of passage for many Swedish young people. Wilma Larson Stewart remembered: "You would have instructions, and then on a Sunday when you were confirmed, the minister asked questions, and you would have an exam in front of the congregation." Ministers questioned young people over catechism and Biblical history. Questions were asked and responses given in the Swedish language. Wilma Larson Stewart wore a white dress and corsage for her confirmation. She remembered it as a very special occasion. The church held Confirmation at the end of the Congregational School each summer.[47]

Christmas represented a special holiday for the Swedes. Wilma Larson Stewart remembered: "The church was beautiful, the candles—[a] big Christmas tree with candles—and then you left early in the morning at 6:00 A.M., . . . early as possible, to go to church." Mrs. Stewart also remembered that her mother prepared many special foods at Christmas: "Well, she baked rye bread, raisin bread, and then there was a bread she rolled out real thin. . . . She made a Swedish drink . . . in a big stone jar, and she sat that by the kitchen stove." Swedes served a special dinner on Christmas Eve and a big family dinner on Christmas Day. The food included sausages, ham, and potatoes, and a soup from the ham broth was always served at noon on the day of Christmas Eve.[48] Alex and Agnes Erickson, who attended Ebenezer Lutheran, recalled that the children had programs at the church on the second day of Christmas, December 26. This was a truncated version of the Christmas celebration in Sweden, the Ericksons recalled, where people celebrated Christmas for an entire month. May Day and Thanksgiving were also highlights of Swedish social life, along with weddings.[49]

Like most ethnic groups, Swedish families often kept boarders—young, unmarried Swedish men who had recently emigrated from Sweden. The Ericksons always had two to four boarders at home. Vera Fisher, also of Swedish descent, recalled that her mother had as many as thirteen boarders. The family rented an extra company house to put up some of these young men. This meant hard work for her mother: "[She prepared] breakfast, then fixed their lunch pail and served supper. And then when they wasn't working, or Sunday, they'd eat three meals a day at our house." For breakfast Mrs. Fisher's mother served bacon, eggs, oatmeal, and biscuits and gravy. When it came to packing lunch pails for the men in the mines, Mrs. Fisher recalled: "Well, it would be lunch meat and butter. She always had pie and cake and that. She did all her own baking."[50]

The Swedes saw themselves as hardworking people, and they certainly were. Wilma Larson Stewart recalled that her mother advised her not to marry a Swede, because Swedish men worked their women too hard. Many Swedish families combined mining with farming for at least part of the family life cycle, and this meant daily chores, which were to be shared by all family members.

Unlike the Swedes, Slovaks in Buxton did not live in one or two specific areas of town. Some Slovaks lived near the Slovak Lutheran Church in West Swede Town and others lived on the west side of Buxton, particularly on West Fourth, West Eighth, and West Ninth Streets. Some Slovakian families also lived on farms outside of Buxton and were known to have "green thumbs" at growing fruit and flowers. But even though Slovaks were not closely identified with a specific part of Buxton, a strong Slovakian community did exist. It was composed of churches, fraternal orders, and parochial schools, and Slovaks assembled regularly for social events. Like the Swedes, the Slovaks also held summer school in Buxton to provide the second generation with instruction and an understanding of the Slovakian language. Joe Re-

Members of the Slovak *Jednota* Fraternal Lodge attending a meeting in Buxton.

barchak recalled that Slovakian school began at 9:00 A.M. and continued until 3:00 P.M. This school was held either in the Slovakian church or in one of the local schoolhouses.[51] Slovaks who had lived in Buxton remembered that the community was cohesive but open to outsiders, and although Slovaks segregated themselves somewhat in private life, they were integrated into the public life of the community.[52]

The religious life of Slovaks centered around two religious groups, the Slovak Lutherans and the Catholics. The Slovak Lutheran Church was located in West Swede Town no more than a block from the Swedish Methodist Church. This was in the northwest quadrant of Buxton. The Catholic Slovaks occasionally held mass at the main YMCA. The mass was held in Latin and English, usually by an Irish priest. Erwin Olsasky recalled: "I walked to Buxton to mass at the YMCA more than once." Slovaks also attended Catholic church at St. Peters in Lovilia, a town about six miles west of Buxton.[53] Once or twice each year Slovaks would bring a Slovakian priest to Buxton and Lovilia to hear confessions in the Slovakian language. Emma Romanco Starks remembered: "They had to hear confession, so the Slovak people that couldn't speak the English language could go to confession, to their own priest." Many former residents remembered that the

Slovakian priest always visited at Easter. Emma Romanco Starks also recalled that some Slovaks were Eastern Orthodox, including her parents. "They were always that," she said, "but after they moved to this country, to this area, why all we had was the Roman Catholic Church, and that's the Church we went to." Sister Maurine Sofranko remembered that an Orthodox priest visited Lovilia once a year to hear confessions from Slovaks.[54]

Like many ethnic groups, the Slovaks belonged to a fraternal order, the *Jednota*. According to Erwin Olsasky, "They'd rent a hall someplace. They used to have . . . meetings in the YMCA building, or either at some house, or they had picnics mostly, maybe out in the country." Charles Lenger recalled that this lodge had five hundred members in the region at one time.[55] *Jednota* provided the usual services of an ethnic mutual-aid society, including insurance. Not all mutual aid among Slovaks was through formal organizations, however. Emma Romanco Starks recalled:

> [Fellow Slovaks needed] $3,000 to buy this farm, and wanted to know if my parents would help them. . . . I happened to be in the kitchen that day. . . . Mother and dad exchanged looks, and they said "Why not. The people need the money, and we've got it." . . . They went to the bank the next day and drew out $3,000 and gave it to these people, . . . and $3,000 would be a million dollars today.[56]

Slovakian social events included picnics, dances, weddings, and holiday celebrations. Dances were held in the homes, usually on farms outside of Buxton. Furniture would be removed from the living and dining rooms and dances would "last about all night, until the next morning." The Slovaks always enjoyed plenty to eat and drink. In fact, both Slovaks and Italians had wine grapes from California shipped by rail to Buxton which they pressed and then fermented for their own wines. They also made their own beer and whiskey.[57]

Slovakian weddings were sometimes three-day affairs. Catholic couples would usually marry on Saturday morning, and some celebrants would have already gathered at the homes of the bride and groom on Friday evening. Some couples married on Fridays, getting an earlier start. The reception would begin immediately after the marriage ceremony and would continue through at least Sunday, breaking up on Monday morning when the men returned to work. These weddings attracted up to two hundred people.[58]

Many former Buxton residents recalled the good food and drink served at Slovakian wedding receptions. Mike Onder exclaimed: "Oh, they had good food. Table from got darn it to yonder." Usually the hosts served stuffed cabbage—rice and meat rolled into a cabbage leaf—and other Slovakian dishes. Food preparation began weeks before the wedding. The hosts always provided something with which to wash down all the food; Erwin Olsasky recalled twenty kegs of beer, twenty gallons of whiskey, and wine being consumed at some Slovakian weddings. According to an Old World custom, the reception was orchestrated by a master of ceremonies, typically a close friend of the family. Toasts would be given and the master of ceremonies would remind the celebrants of the significance of the occasion.[59]

The music and dancing also followed Old World customs. Violins were the lead instruments in the band. At some ceremonies, older women would encircle the bride on the dance floor:

> then they'd all join hands and put the bride in the middle, and they'd go in a circle and sing to her. . . . They'd sing. They'd stick with it until she cried . . . sort of sad songs, you know, after marriage and this and that. . . . Well, what she [would] run into. You don't know what the future is and all that.[60]

Slovaks did not always observe this custom, but frequently they did hold money dances. Men paid to dance with the bride, and sometimes this money would be pinned on the bride's dress. Wedding celebrations commonly lasted all night. People slept wherever they could, and the music and dancing would begin again the next day, following a late breakfast. The bride and groom remained for the entire reception, and the men, including the groom, then returned to work on Monday morning. Slovakian newlyweds did not have honeymoons.

Christmas and Easter were also special times for Slovaks. The meal on Christmas Eve was a traditional one:

> On Christmas Eve, we had a very special kind of dinner. It was a ceremony we had. I remember . . . mushroom soup. That was one thing, and there were other particular dishes we had to have. My grandfather would read at the table. . . . He'd read from Scripture. We had unleavened bread. . . . First there were prayers, followed by a serving of a thin wafer of unleavened bread . . . which was dipped in honey. . . . There was a sweet- and-sour mushroom soup.[61]

Many families attended mass at midnight on Christmas Eve and again on Christmas morning. Slovaks exchanged gifts on Christmas Day and also enjoyed big meals. Erwin Olsasky recalled that food was on the table from Christmas Eve throughout Christmas Day. Men made special drinks for Christmas and joined in several toasts. Slovaks even extended the spirit of the season to their livestock. Erwin Olsasky recalled that Slovakian farmers gave special care to their livestock at Christmas and fed them special provisions.[62]

Easter was also special, with Slovakian priests visiting Buxton and Old World customs again being observed:

> Mother would make Easter eggs, and she made them like they did in the old country. She'd boil the eggs, then she had a little stick that she would dip in wax and make figures on the egg, and then she would dye the egg, and then to get that wax off an egg . . . she would heat the egg.[63]

Ham, sauerkraut, mushroom soup, and Easter eggs were traditional Easter foods. Slovakian housewives always served the *kolac*, the national pastry of the Slovaks and Czechs for holidays. Most Slovaks recalled liking the ones with poppy seeds in the center.

Although former residents of Buxton, both black and white, repeatedly asserted that blacks experienced no racial discrimination there, it is necessary to examine the matter of racial discrimination from several perspectives. While it seems that overt racial discrimination did not exist in Buxton, it is possible for discrimination to take subtle, or even hidden, forms. Because of the size and diversity of Consolidation's coal operation, as well as the size and diversity of the company town, there were numerous levels at which racial and ethnic discrimination might have been practiced by company personnel. Aboveground, discrimination might have been practiced in regard to housing assignments, the treatment of customers at the company store, employment in areas other than coal mining, and treatment at public facilities. Belowground, in the mining process itself, discrimination could have been present in the assignment of jobs or rooms, differential wage rates, and the segregation of transportation to and from work. The mining work force at Buxton included miners, company personnel, and independent coal operators. Company personnel included management positions (e.g., superintendents, pit bosses, and foremen), clerks, manual workers belowground (e.g., trappers, mule drivers, timbermen, and shot firers), and manual workers aboveground (e.g., weighmen, hoisting engineers, and night watchmen).

Some racial differences existed in the Buxton mining work force in 1905 (Table 6.3). The company hired more whites than blacks as company personnel and this racial difference was greater in 1905 than in 1915. If one were to conclude, however, that this is evidence for favoritism shown whites in 1905, than one must explain why whites subsequently turned their backs on this favoritism and became more concentrated as coal diggers or miners by 1915.[64]

Job assignments in coal production at Buxton in 1915 revealed no significant racial differences. Virtually all workers were miners in 1915. Whites were more likely than blacks to be company personnel, but the number of these workers is small and the racial differences are small. It seems that there was no clearcut racial stratification of the mining work force at Buxton in 1915. The one exception is that the superintendent of the mines at Buxton was always a native-born male, indicating that there was racial and ethnic stratification at the very top of the work hierarchy.[65]

The percentage distributions of foreign-born workers by occupational group in coal production at Buxton for 1905 and 1915 are given in Tables 6.4 and 6.5. Workers from the British Isles and Sweden were represented among company personnel, especially in 1905, but this was limited to only a few workers, and both Swedes and those born in the British Isles became more concentrated as miners between 1905

Table 6.3. Percentage Distribution and Number of Black and White Workers by Occupational Group in Coal Production, Buxton, 1905 and 1915

Occupational Group	1905				1915			
	Whites		Blacks		Whites		Blacks	
Miner	86.1%	(254)	96.5%	(906)	96.0%	(528)	98.3%	(514)
Company personnel (management)	3.4	(10)	0.1	(5)	0.9	(5)	0.4	(2)
Company personnel (clerical)	0.7	(2)	0.1	(1)	0.4	(2)	0.0	(0)
Company personnel (manual, belowground)	3.7	(11)	2.5	(23)	1.8	(10)	0.6	(3)
Company personnel (manual, aboveground)	4.4	(13)	0.4	(4)	0.9	(5)	0.8	(4)
Coal operator	1.7	(5)	0.4	(4)	0.0	(0)	0.0	(0)
Total	100.0	(295)	100.0	(943)	100.0	(550)	100.1	(523)

Source: Iowa, Census of Iowa, 1905 and 1915, Manuscript Population Schedules for Bluff Creek Township, Monroe County.

Note: Retired miners and counselors in mines have been excluded (N = 9). Percentages may not sum to 100 because of rounding.

Table 6.4. Percentage Distribution and Number of Foreign-Born Workers by Occupational Group in Coal Production, Buxton, 1905

Occupational Group	Place of Birth: Sweden		Slovakia		British Isles	
Miner	97.2%	(69)	100.0%	(22)	91.9%	(34)
Company personnel (management)	0.0	(0)	0.0	(0)	5.4	(2)
Company personnel (clerical)	0.0	(0)	0.0	(0)	0.0	(0)
Company personnel (manual, belowground)	0.0	(0)	0.0	(0)	0.0	(0)
Company personnel (manual, aboveground)	2.8	(2)	0.0	(0)	2.7	(1)
Coal operator	0.0	(0)	0.0	(0)	0.0	(0)
Total	100.0	(71)	100.0	(22)	100.0	(37)

Source: Iowa, Census of Iowa, 1905, Manuscript Population Schedules for Bluff Creek Township, Monroe County.

Note: The labor force in coal production also included 3 Italians.

Table 6.5. Percentage Distribution and Number of Foreign-Born Workers by Occupational Group in Coal Production, Buxton, 1915

Occupational Group	Place of Birth: Sweden		Slovakia		British Isles	
Miner	100.0%	(43)	100.0%	(102)	92.5%	(37)
Company personnel (management)	0.0	(0)	0.0	(0)	5.0	(2)
Company personnel (clerical)	0.0	(0)	0.0	(0)	0.0	(0)
Company personnel (manual, belowground)	0.0	(0)	0.0	(0)	0.0	(0)
Company personnel (manual, aboveground)	0.0	(0)	0.0	(0)	2.5	(1)
Coal operator	0.0	(0)	0.0	(0)	0.0	(0)
Total	100.0	(43)	100.0	(102)	100.0	(40)

Source: Iowa, Census of Iowa, 1915, Manuscript Population Schedules for Bluff Creek Township, Monroe County.

Note: By 1915, 52 Italians worked as miners in Buxton.

and 1915. The number of foreign-born Italians in coal production at Buxton increased from three workers in 1905 to fifty-two in 1915, and all Italians were miners.[66]

The same trends are found for the second generation of these ethnic groups in Buxton. The second generation is defined by father's birthplace, that is, having a father born in the British Isles, Sweden, or Austria-Hungary (Slovakia). Members of the second generation from the British Isles were more likely to be company personnel in 1905 and 1915 than the second generation from Sweden, Slovakia, or

Italy. However, there were never more than six second-generation workers from the British Isles in company work.[67]

Company personnel in coal production at Buxton tended to be white and native-born. Moreover, the foreign-born parentage among company personnel from the British Isles tended to be from Great Britain rather than Ireland. This suggests some semblance of a cultural division of labor between miners and company personnel along racial and ethnic lines. The number of company personnel was always so small, however, that it is difficult to draw a conclusion about the racial and ethnic stratification of these jobs. Company personnel were only 5.6 percent of the total work force in coal production at Buxton in 1905 and 3 percent in 1915.[68]

Between 1905 and 1915, the number of white miners at Buxton grew from 254 to 528, while the number of black miners fell from 906 to 514. Whites were displacing blacks as miners over this decade. White miners increased in number by 274, and 48 percent of this increase was due to the growth in the numbers of Slovak (80) and Italian (52) miners at Buxton over this same decade. In the preceding decade, the number of black miners grew from 254 men in 1895 at Muchakinock to 906 in 1905 at Buxton, a 257 percent increase. During this same period, the number of white miners declined from 354 to 254 men, a 28 percent decrease. In the subsequent decade, however, the situation was reversed. White miners displaced black miners between 1905 and 1915, increasing by 108 percent while blacks declined by 43 percent.[69]

Several factors might account for these changes. First, the total number of blacks in Buxton had decreased by 1915. Secondly, the mining profession contained several inherent conditions that encouraged men to become miners rather than company personnel. First, it was generally understood that miners could make more money than company personnel, and, second, miners experienced more independence in their work. (See chapter 3 for a full discussion of these conditions.)

In addition to Consolidation's policy of equal pay for equal work, workers also had the backing of the United Mine Workers of America in regard to equal-wage policies, regardless of workers' race or ethnic background. The only wage differentials permitted by the UMW were by occupational category. The implementation of this equal-wage policy was confirmed in the oral histories of former miners at Buxton.

The inside of the blacksmith shop in Buxton. The message on the side of the stove, "Gone but not forgotten," perhaps pays tribute to a fatally injured miner, J. C. Carlson, who died in 1903. (*Courtesy of Herman Brooks*)

Mike Onder reported, for example, that black and white miners received the same pay from Consolidation, and he remembered no instance of differential wages by race or ethnic backgrounds. Earl Smith reported the same thing. Indeed, most former residents of Buxton, both blacks and whites, recalled that the company treated blacks and whites the same.[70]

If not in wages, then discrimination could have occurred in room assignments. Perhaps members of one group were, on the average, given rooms with dirtier coal or kept waiting longer for their assignment to a new rooms. Such discrimination would be reflected in differential wages. No such discrimination was found, however, at least with respect to race. A black former miner, Jacob Brown, commented on such possible favoritism by saying, "It wasn't allowed to be." He then added: "Course if the foreman kinda got something against you, they [might] give you a bad one [room] or one with dirty coal." Brown's statement suggests some discrimination, but not by race or ethnic background.[71]

Miners began their workdays in Buxton by walking to the train depot to catch the miners' train for the ride to the mines. There is no evidence of segregated seating on these trains. The trains contained no separate cars for black miners, nor did single cars contain white and black sections. Former miners recounted that miners sat where they wished, and no one reported any resistance among miners to an integrated train. Harvey Lewis recalled that his father, a Welsh miner, "rode the miners' train with the blacks. . . . There were probably four blacks to one white on the miners' train. There never was any trouble." Whatever racism or ethnocentrism might have been present in Buxton was not made manifest in discrimination at work. Rather, former residents reported that relations among the racial and ethnic groups in the workplace were good.[72]

While work activity belowground provides one area for possible discrimination, life aboveground presents another. Intergroup relations within the community itself need to be examined. This will be done in both the areas of public and private life. Table 6.6 shows the occupational composition of blacks and whites in Buxton in 1905 and 1915. In 1905 black workers in Buxton were clearly concentrated in coal mining, and, as seen earlier, over 95 percent of blacks in coal production worked as miners. Another 13 percent of Buxton's black workers were either service workers or laborers, including domestics. Whites were also concentrated in three occupational groups in 1905:

Table 6.6. Percentage Distribution and Number of Black and White Workers by Occupational Group, Buxton, 1905 and 1915

Occupational Group	1905				1915			
	Whites		Blacks		Whites		Blacks	
Professional and semiprofessional	3.7%	(30)	1.5%	(18)	1.8%	(16)	2.7%	(19)
Business	6.7	(55)	1.4	(17)	5.5	(50)	2.5	(18)
Artisan	5.0	(40)	3.5	(42)	2.2	(20)	4.5	(32)
Coal production	36.1	(295)	78.6	(939)	60.2	(550)	73.5	(523)
Transportation and communication	3.1	(25)	1.0	(12)	5.5	(50)	1.7	(12)
Service work	2.8	(23)	4.7	(56)	1.0	(9)	3.5	(25)
Laborer (including domestic)	17.4	(142)	8.4	(100)	3.5	(32)	8.9	(63)
Agriculture	25.3	(207)	0.8	(10)	20.5	(187)	2.8	(20)
Total	100.1	(817)	99.9	(1,194)	100.2	(914)	100.1	(712)

Source: Iowa, Census of Iowa, 1905 and 1915, Manuscript Population Schedules for Bluff Creek Township, Monroe County.

Note: Workers in miscellaneous jobs, the disabled, retired, and students have been excluded (N = 91). Percentages may not sum to 100 because of rounding.

miners, laborers, and farmers and farmhands. These three occupational groups accounted for 79 percent of the white workers in Buxton in 1905. There was, however, a greater percentage of white than black workers in the following areas: professional and semiprofessional, business, artisan, and transportation and communication (much of it railroad work). There was also a smaller percentage of whites in service work. These ratios imply some measure of racial stratification to the disadvantage of blacks. At the same time, however, whites were more concentrated than blacks in the class of common labor, and this has the reverse implication. A much greater percentage of whites than blacks were in agriculture, which seems to have been virtually a white monopoly in 1905.[73]

By 1915 the percentage of whites in agriculture and common labor had declined, as had the percentages of whites in the professions and business. White workers in mining rose from 36 to 60 percent of the total white work force in the same decade, and 96 percent of these were miners (Table 6.3). On the other hand, the percentage of black workers in mining fell slightly, while the number of blacks in mining fell sharply. There was a slight rise in the percentage and number of blacks in the professions, in business, and in agriculture.[74]

The occupational composition of blacks in Buxton was very different from that in the nation as a whole at this time. In 1910, to illustrate, over 50 percent of the employed blacks in the United States were in farm work, and another 21.3 percent were service workers. Two sociologists, Sterling Spero and Abram Harris, summarized the general situation of blacks: "The emancipation of the slaves in the perspective of the labor movement produced the following results. The major portion of the Negro labor supply was shunted away from the labor movement and industrial employment into agriculture and domestic service."[75]

With the passage of Jim Crow legislation in the 1880s and 1890s, blacks were disfranchised in southern states, segregated in public accommodations and on common carriers, and hindered in their access to due process of law. With the *Plessy* vs. *Ferguson* decision of the United States Supreme Court in 1896, they were also segregated in public education. Blacks were made powerless, in a word, and set up as targets for white hostility, which helped check any black challenge to the racial caste system.[76]

This was not, however, the history of blacks in Buxton. Blacks there were not concentrated in agriculture and service work, for no more than five percent of the black workers in Buxton were in these two occupational groups. Blacks in Buxton were not shunted away from industrial work and the labor movement. Buxton did not contain a racially split labor market, so evident in the South, and Buxton had no Jim Crow laws. In a word, there was no racial caste system in Buxton.

The foreign-born population represents another group in Buxton that offers the possibility of occupational discrimination. The percentage distributions of foreign-born workers by occupational group in 1905 and 1915 at Buxton are shown in Tables 6.7 and 6.8. In both

Table 6.7. Percentage Distribution and Number of Foreign-Born Workers by Occupational Group, Buxton, 1905

	Place of Birth					
Occupational Group	Sweden		Slovakia		British Isles	
Professional and semiprofessional	0.0%	(0)	0.0%	(0)	3.4%	(2)
Business	5.1	(6)	3.2	(1)	1.7	(1)
Artisan	4.2	(5)	0.0	(0)	1.7	(1)
Coal production	60.2	(71)	71.0	(22)	64.4	(38)
Transportation and communication	0.8	(1)	0.0	(0)	0.0	(0)
Service work	3.4	(4)	0.0	(0)	1.7	(1)
Laborer (including domestic)	22.9	(27)	25.8	(8)	8.5	(5)
Agriculture	3.4	(4)	0.0	(0)	18.6	(11)
Total	100.0	(118)	100.0	(31)	100.0	(59)

Source: Iowa, Census of Iowa, 1905, Manuscript Population Schedules for Bluff Creek Township, Monroe County.

Note: Workers in miscellaneous jobs, the disabled, retired, and students have been excluded ($N = 4$). Percentage may not sum to 100 because of rounding.

Table 6.8. Percentage Distribution and Number of Foreign-Born Workers by Occupational Group, Buxton, 1915

	Place of Birth					
Occupational Group	Sweden		Slovakia		British Isles	
Professional and semiprofessional	0.0%	(0)	0.0%	(0)	0.0%	(0)
Business	5.6	(3)	1.8	(2)	4.1	(2)
Artisan	3.7	(2)	0.9	(1)	0.0	(0)
Coal production	79.6	(43)	89.5	(102)	81.6	(40)
Transportation and communication	1.9	(1)	0.0	(0)	2.0	(1)
Service work	1.9	(1)	0.9	(1)	2.0	(1)
Laborer (including domestic)	1.9	(1)	0.0	(0)	2.0	(1)
Agriculture	5.6	(3)	7.0	(8)	8.2	(4)
Total	100.2	(54)	100.1	(114)	99.9	(49)

Source: Iowa, Census of Iowa, 1915, Manuscript Population Schedules for Bluff Creek Township, Monroe County.

Note: Workers in miscellaneous jobs, the disabled, retired, and students have been exluded ($N = 8$). Percentages may not sum to 100 because of rounding.

1905 and 1915, foreign-born workers at Buxton were concentrated in mining. While 36 percent of all white workers at Buxton in 1905 were in coal production, 60 to 71 percent of the foreign-born workers were found in that single occupational group. The concentration of the foreign-born in mining in 1905 is nearly equal to that of blacks. A higher percentage of the foreign-born were common laborers than blacks or whites as a whole.[77]

By 1915, however, this picture had changed, as the foreign-born became more concentrated than blacks in coal production. As previously discussed, the foreign-born and their second generation were important elements in the displacement of black miners by white miners between 1905 and 1915 in Buxton. The number of foreign-born Slovaks in coal production increased from 22 in 1905 to 102 in 1915. The number of second-generation Slovaks in mining also increased, from 25 in 1905 to 123 in 1915. The same trend is found for both generations from the British Isles. For this ethnic category, it is especially the growth in the number of the second generation in mining (69 workers in 1905 to 96 in 1915) that accounts for the trend. The number of foreign-born and second-generation Italians working as miners in Buxton also increased between 1905 and 1915. The state census of 1905 included only 3 Italian miners at Buxton, while the census of 1915 included 52 Italian-born males and 53 second-generation Italian-Americans working as miners there.[78] The number of foreign-born Swedes in coal production at Buxton fell between 1905 and 1915 by 39 percent, from 71 workers to 43. Second-generation Swedes in mining also declined from 87 workers to 69, a 21 percent decrease. Thus the two oldest ethnic groups in Muchakinock and Buxton, blacks and Swedes, were being displaced in mining by newer immigrants between 1905 and 1915.[79]

In regard to race and ethnic relations in public life, the issues of residential patterns and access to education, public accommodations, and services are also important. Most housing in Buxton was company owned, so racial segregation or integration in residence would have been a reflection of company policy. When asked about segregation in housing at Buxton, Jacob Brown replied: "Well, yes and no. Now, when people came there . . . they moved generally where they could get a house. 'Course after they got there, they could move around there sometimes. But you could go anywhere you wanted to and never be bothered, as long as you acted right."[80] As Brown mentioned, when families moved to Buxton they were assigned company housing on a first come, first served basis. The impression is that company housing in Buxton was racially integrated at first, but that people could later relocate and possibly segregate themselves. The Swedes were an important exception, for many built their own houses, concentrating in Buxton's two Swede Towns. Both Swede Towns and areas immediately outside Buxton proper were predominantly white.

Most of Buxton proper is shown in Figure 1, with the percentages of black household heads by streets in 1910. Buxton was divided by Main Street into east and west sections. Federal census takers contacted households along the east and west sections of numbered streets, either moving up one side of the street and down the other or in a criss-cross pattern. The results shown in Figure 1 are the percentages of black household heads by the east and west sections of the numbered streets. Sixty-four percent of the household heads in Buxton proper in 1910 were black and 36 percent were white. As one proceeds south, from First Street and downtown, the percentage of black household heads generally increases. Starting with Third Street, the percentage of black household heads begins to exceed the percentage for Buxton proper (64 percent), that is, the area shown in Figure 1 plus East and West Swede Town.[81]

The concentration of black household heads actually indicates a hopscotch pattern. Black household heads predominated east of Main Street from Third to Sixth Street, and then the concentration jumped over to West Sixth and Seventh Streets. On Eighth and Ninth Streets, the percentages of black household heads predominated again on the east side of Main Street. On Tenth, Eleventh, and Thirteenth Streets, there were only small social differences between the west and east sides. Black household heads predominated on West Twelfth and again on West Fourteenth.[82] Below Fourteenth Street, to the south and out of Buxton proper, virtually all households were white. This is to be expected, because agriculture was a white monopoly in the Buxton area. Whites also predominated in East and West Swede Town, off the map in Figure 1. About the Swede Towns, Odessa Booker recalled, "Swede Town was nothing but Swedes." That is not absolutely true, for many Slovaks and others also lived in the so-called Swede Towns.[83]

Another area of public life that should be examined for ethnic and racial discrimination is education. There were three public grade schools in Buxton, one in East Swede Town and two in Buxton proper.

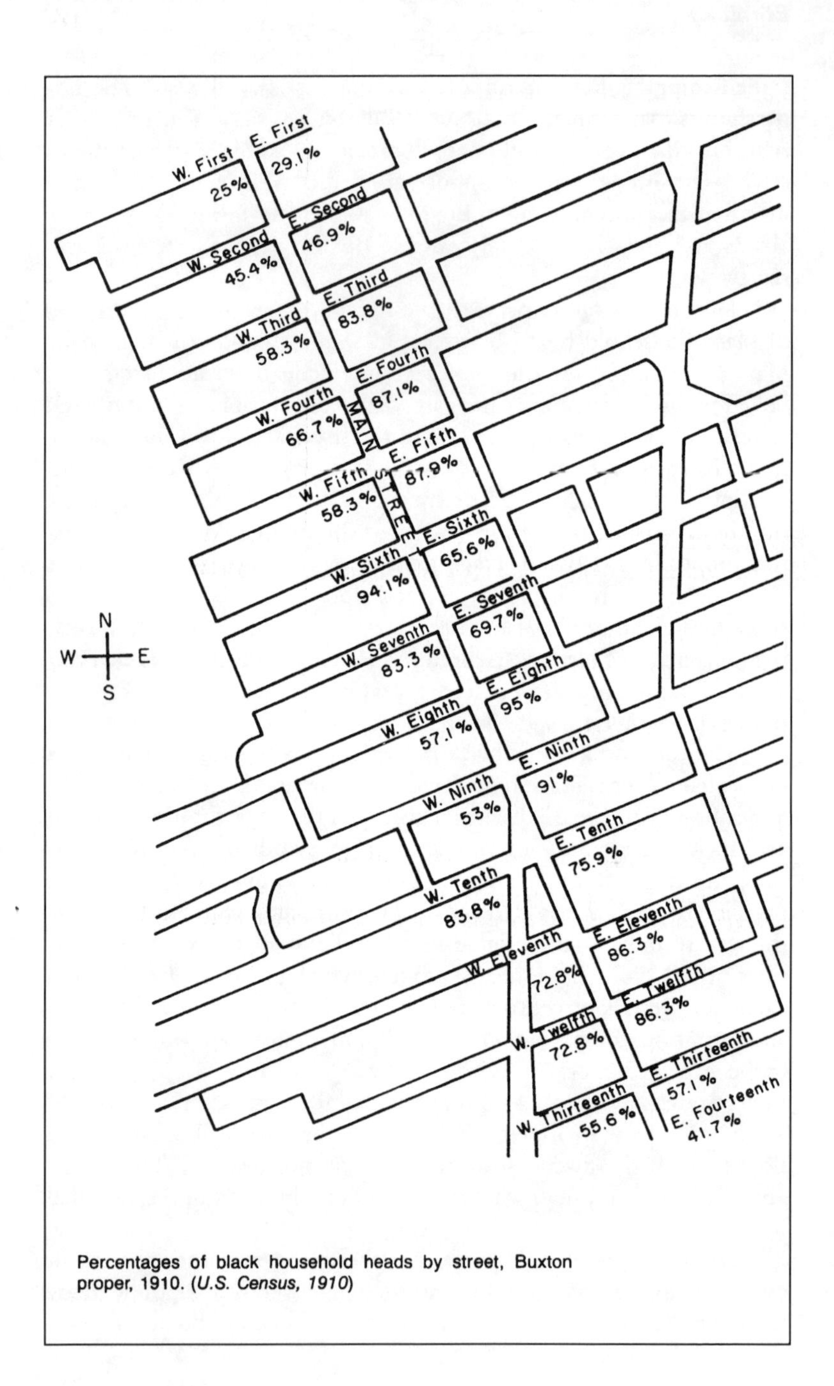

Percentages of black household heads by street, Buxton proper, 1910. (*U.S. Census, 1910*)

Buxton also briefly had a high school. It was built in 1906 but was destroyed a fire a year later and never rebuilt. The grade schools were racially integrated with the exception of the school in East Swede Town, which was mostly white because it served a predominantly white neighborhood.[84]

Many former Buxton residents reported attending integrated schools. Robert Wheels, for example, attended the Fifth Street School through the eighth grade, and he recalled that there were both black and white students and teachers. Helen Clay remembered that the schools "were mixed schools" and that "some [teachers] were white and some was colored." Gertrude Stokes stated: "We had colored and white teachers." When asked about racial segregation in Buxton, Dorothy Collier commented, "I don't ever, even in school, I don't remember any segregation in the school."[85] There was little mention of racial ill will in the schools. Earl Smith noted, however, that "I went to school at Eleventh Street one time. I only went there for about a week. Them kids were too rough for me. . . . Them was all colored teachers at Eleventh Street School, most[ly] colored children." On the other hand, Jacob Brown recalled having white teachers at the same school.[86]

Many former residents remembered the school in East Swede Town as mostly white, in regard to teachers as well as students. Charles Lenger started school at the Fifth Street School in Buxton proper and then transferred to East Swede Town, "for some reason, on account of my friends I played with, you know." He recalled that this school was mostly white. Wilma Larson Stewart remembered a few black classmates at the East Swede Town school, as did Agnes Erickson; Jacob Brown recalled that in regard to school in East Swede Town, "there were some colored that went over there, the ones that lived close enough in that district." Alex and Agnes Erickson remembered that in Buxton the Swedes "had a private white school for a while, then it all became a public school." The reason for the change, according to the Ericksons, was that "they wanted the white kids to . . . mix with the blacks." As to public school, however, Agnes Erickson recalled: "Some of the West [Swede] Town kids would walk through all of Buxton up to the East Swede Town. . . . One of the schools right in the center of town, it was nearly all colored, and they didn't want to go with them, so they'd come up to the other Swede school." In September 1912 the *Bystander* commented that the teachers at the East Swede Town school were all white.[87]

Social life in Buxton was frequently punctuated with educational events, such as spelling bees, presentations of papers by local residents, and guest speakers from all over the nation, including scholars from the Tuskegee Institute and prominent black businessmen and women. Graduation exercises were special occasions in Buxton and were regularly reported in the *Bystander*:

Monday evening in the YMCA auditorium a magnificent audience witnessed the commencement of the 8th grade pupils of the Buxton schools. The address to the graduates was given by Dr. E. A. Carter and was considered by all to be quite wholesome and full of helpful instruction. Miss Myrta Harlow, County Supt. . . . presented the diplomas, prefaced by some helpful remarks. Miss Harlow is much pleased with the work of the Buxton teachers and spoke of them in very high terms.[88]

By all accounts, then, blacks had equal access to public education in Buxton, although the racial composition of the schools varied. This reflects the varied racial composition of certain neighborhoods and the particular situation in East Swede Town. Buxton residents took education seriously. It was not uncommon, however, for children to be withdrawn from school at an early age to begin work in the mines. The concern for education manifested itself in public meetings about the schools, and former teachers and students recalled a basic education of some quality in Buxton. Moreover, educational events were woven thoroughly into the community's larger civic and social life.

In examining public accommodations from the standpoint of discrimination, it should be remembered that Buxton existed in the era of the Jim Crow laws. Yet almost all former residents interviewed remembered that blacks had equal access to all public accommodations and facilities there. Earl Smith recalled that there was no enforced segregation of the races. When asked the reason for this, Smith replied: "In those days everybody was equal as far as that company was concerned; they was all the same." Unrestricted access by both races to all public places and accommodations in Buxton, from the company store to restaurants and hotels, was confirmed in other oral histories. Vaeletta Fields and Susie Robinson both recalled that blacks and whites roller skated together. Both were specifically asked if blacks and whites skated on different nights, and both reported that they skated together on the same nights. Clifford Lewis remembered that at this time blacks were not served with whites at restaurants in Albia, the county seat, a town no more than ten miles from Buxton, but no former residents remembered any segregation at the Buxton company store, including its soda fountain. Archie Allison recalled that blacks had full access to all public accommodations in Buxton while blacks were stepping off the sidewalks for whites in nearby Missouri. Earl Smith reported that seating at the ballpark to watch the Buxton Wonders was open to all regardless of race. Officials handled seating on a first come, first served basis, and no black and white sections existed in the stands.[89]

Movies shown at the Buxton YMCA were also integrated. Gertrude Stokes remembered just as many whites as blacks at the movies on the second floor of the YMCA building. Many recalled a mixed seating pattern at the movies. Clara Jones and Mike Onder remembered some occasional voluntary segregation at the movies. Onder recalled: "They kind of mixed once in a while but not too much. . . . Most of [the] colored [sat] on the right side. They didn't have to, but . . . most of them did it. The black was on the right side, and the white on the left side."[90]

Still, racial integration can occur in public life while racial separation remains the norm in private life. Even public events that imply an intimacy among participants are often segregated. Dances would be this kind of social event, along with church services and club meetings. Archie Allison recalled that dances were largely segregated by race in Buxton, but access was always possible under sponsorship. "I've watched the black dances," he said, "but I never went in, and I never was invited in." Allison recalled that blacks were sometimes invited to watch white dances, but "you had to be invited as a guest, otherwise there could be the damnest fights you ever seen." Nellie King recalled that black dances were virtually all-black events, although white bands played at times. Gertrude Stokes commented on the racial separation in private life: "Yeah, they were friendly with one another. When we had our parties and things, we didn't mix like that. . . . They were all colored, and I guess the whites had theirs, . . . and I guess they had their good times too."[91]

Children were different, however, for they were less observant of any color line in private life. Oliver Burkett recalled that his best friend was a white boy whose family lived across the street. Naomi Ambey and Helen Clay remembered having white playmates. Helen Clay commented: "We had white playmates, but we never did have trouble that I know of." Clyde Wright remembered white friends often staying all night at his house, and Elmer Buford recalled eating with families of white friends, skating and sledding with friends of both races in the winter. He added: "We used to play that, oh, leap frog, we used to play a lot of marbles, we'd shoot marbles . . . we'd play keeps."[92]

Dorothy Collier remembered that her father had white friends, team members of the Buxton Wonders, and he often brought them home. She also recalled her mother having friends of both races, and they got together for quilting bees:

> The white and colored, . . . Mama had both for friends. . . . Ladies would come over and visit. . . . In the winter, the women would put a quilt in a frame, and then they'd all, I guess the neighbor friends, would come in and they'd quilt that quilt.[93]

Intermarriage happened infrequently in Buxton, but it did happen. Nellie King, a white woman, married a black man. Mary Allison recalled: "One of my best girlfriends, her mother was married, and her mother was white, and her father was a colored man." When asked how intermarriage was accepted at the time, she replied, "Well, not very well." Dorothy Collier responded to a question about interracial dating and marriage in Buxton by saying that neither happened frequently. The color line in Buxton was not drawn in public life, nor was it enforced in any formal authority; rather racial segregation was a private matter, done in private life, more by some than others. Albert and Vera Fisher remarked: "Nothing for you to work one room [in the mines] and a colored guy to work the next room. We got along all right. You just didn't associate with them after you got away from [work]. Spoke to them on the street, but that was it."[94]

To get a better idea of the social distance between the races in Buxton, we turn to Ada Morgan and the social distance between the classes there. Ada Morgan's father became superintendent of the Buxton mines in 1909, when Ada was a teenager. As an only child, Ada lived with her parents in the roomy house provided for the superintendent and his family. A large meadow separated the superintendent's house from Buxton proper. The Morgan family's circle of acquaintances remained extremely limited:

> Well, I enjoyed the few people that we became acquainted with. Some of the men in the office, and none were married at the time, and there was a woman in the office, we were good friends, and the manager of the store, his wife was a sister of Mrs. Buxton. They were very pleasant to be with, and the doctor

and his wife. The white doctor and his wife were from Albia . . . and another doctor and wife from Albia came, and we were very friendly with them.[95]

When asked if she had known any black families, she replied: "No, . . . [but] I didn't know any white miners' families either. That meadow seemed to separate us." The metaphor of a meadow conveys a far greater sense of social distance between classes in Buxton than between races or ethnic groups.[96]

In the nineteenth century, great changes occurred in American society. In the Northeast, the nation's industrial base took shape, and Americans moved west in large numbers. These changes largely bypassed black people, however, resulting in their divergence from the historical course of the larger nation. Blacks did not resettle on the frontier in great numbers, nor did they migrate north for industrial work until much later. Instead, they remained in the South, where they were forced into a caste of propertyless agrarian workers and domestic servants. They also faced the reality of Jim Crow laws and Black Codes, further eroding any opportunity to advance themselves socially or economically.

The history of blacks in Buxton must be put in a different context. In 1910, 90 percent of American blacks lived in the South, most in the rural South. The Consolidation Coal Company first brought blacks to Muchakinock in 1881; this population continued to grow and later moved to Buxton. Most of the black migrants came from the South, particularly from Charlottesville and Staunton, Virginia. The black population of both Charlottesville and surrounding Albemarle County declined in this period, as the white population increased.[97] Thus Buxton's blacks had left the South and what it represented before the major black exodus later in the century.

Consolidation first brought blacks to Muchakinock as an alternative to the striking white labor force, and a paternalistic relationship formed between the management and the black workers at Muchakinock and Buxton. Bessie Lewis recalled how the company enforced racial harmony. She related that in the case of dissidents, "They couldn't live there very long 'cause they'd fix it so they wouldn't be there."[98] This relationship between management and labor lasted even after the unionization of Buxton miners and was in fact reinforced in joint wage agreements with the United Mine Workers of America.

Over time, however, white miners—mainly first- and second-generation Slovaks and Italians—replaced black miners at Buxton, reversing the earlier pattern of blacks moving into jobs left by whites. For a time, however, blacks had found a niche with the Consolidation Coal Company, one that went far beyond their work.

But Buxton consisted of more than black workers. Both native-born and foreign-born whites were always a substantial part of the community's population. Swedes and Slovaks made up distinct ethnic communities within Buxton and were concentrated in the ranks of miners. Some cultural variations appeared between the two groups. As a rule, Swedes lived in one of two specific parts of town, while Slovaks tended to be more scattered. Moreover, Swedes tended to own their own homes.

Given the fact that Buxton remained a biracial, multiethnic community throughout its existence, there was always the opportunity for both racial and ethnic discrimination, but this does not seem to have occurred. No evidence of racial discrimination appeared in the community—in housing, education, or public accommodations—or belowground in the mining work itself. For blacks in particular, Buxton was indeed a very special place.

7

BUXTON AND HAYDOCK: THE FINAL YEARS

OURTEEN years after its founding, the death knell began to sound for the community of Buxton. In 1914 a series of events took place that produced a leveling off and then a decline in the town's population and business life. In that year Consolidation opened Mine No. 18 ten miles west of Buxton and created the coal camp of Consol. Some Buxton families relocated there immediately. The same year an explosion took place at Mine No. 12 that permanently closed the mine. Although the company later reopened other mines nearby, the shutdown of No. 12 resulted in some families leaving the area. Because of the mine shutdown and the results of earlier labor disputes, Buxton miners worked fewer days in 1914 than previously. This loss of work also prompted some families to leave. From 1914 on, the *Bystander* carried periodic reports of people moving away. During 1916 and 1917, Buxton experienced a series of fires that further hastened its demise. Then in 1918 the major exodus began. In that year Consolidation began moving families to another new camp, Haydock, and within three years the company had relocated all its employees there. Those who did not wish to move to Haydock left the employment of Consolidation and relocated elsewhere. Even as early as 1919, the *Bystander* editor described Buxton as a deserted town. Four years later, Buxton ceased to exist.

Although 1914 would be a turning point in the life of Buxton, in January of that year nothing seemed more remote. Buxton had always had some population turnover, but there had been no significant exodus from the community to this time. In fact, throughout 1912 and 1913 Buxton had continued to attract additional residents.

Among the newcomers were Mrs. I. M. Mardis and her two daughters, Ada and Pearl, who arrived in 1912 to make their home with two other Mardis children, Gussie and William. William worked as a miner and Gussie worked as the supervisor of the jewelry department in the company store. Pearl and Ada immediately went to work as dressmakers. With all four children employed, the Mardis family must have lived well. No doubt Mrs. Mardis and her daughters had been attracted to Buxton for the same reasons as many others, to join family members already there and to settle in an economically prosperous community. Apparently many others felt the same, because in September 1912 the *Bystander* announced that "Quite a number of new faces are coming into Buxton of late. Mostly colored faces too."[1] Throughout 1912 and 1913, Buxton residents continued to organize new clubs and businesses. In May 1912 the *Bystander* mentioned the Buxton Civil League for the first time; the same month, residents organized a new literary society. In November 1912, D. T. Whitaker arrived to start a newspaper, the *Buxton Leader*. The following year, Buxton viewed talking pictures for the first time at the Coopertown Opera House.[2]

From 1914 on, however, a certain ambiguity characterized the reports on Buxton. On the one hand, the *Bystander* reported on Buxton's business and institutional life much as it had done earlier. Buxton continued to be known as a black community, and Buxton's business life remained dominated by black residents. Reports in the *Bystander* indicate that all churches continued to hold regular services, revival meetings, and social events. The YMCAs continued to sponsor events, including night school and sporting events. On the other hand, however, news items on Buxton appeared less and less frequently. When articles did appear, they often reported on families who were moving away.

The first evidence of economic difficulty came in the form of miners' layoffs. Beginning in December 1913, Buxton miners experienced periodic unemployment that lasted until mid-1914. Although the *Bystander* did not state the reason for the layoffs, they probably stemmed from a lack of railroad contracts, which in turn reflected a decreased demand for coal by the railroad. In January 1914 the *Bystander* reported: "There has not been much work in Buxton for the last month, but everybody seems to be in good cheer."[3] Two months later, in addition to the general layoffs, an explosion in Mine No. 12 idled even more miners. The *Bystander* reported that the explosion had caused severe damage, and that the shaft had fallen in. Because of the extensive damage, the company decided not to reopen the mine, an action that threw three hundred men out of work. It is not known how long these men were unemployed before they were reassigned to other Consolidation mines. Apparently most miners did not go back to work full-time until June. These conditions led the *Bystander* to observe: "This has been a very hard year on us. This spring lots of men have been out of work, so that makes it hard on all of our societies and churches. Lots of our best men and women are leaving, going to the cities to work." Several weeks later the *Bystander* noted that work had still not returned to normal: "We are still looking for better things in Buxton. We think work will boom after a while." Two weeks later conditions had apparently improved, as the *Bystander* reported that miners were working every day. The extended layoff, the longest in Buxton's history, was highly unusual in a railroad mine.[4]

Although it is difficult to pinpoint which people began leaving Buxton in 1914, interviews and newspaper accounts indicate that black families were often the first to depart. While some blacks had purchased land in the area (usually acreages of around fifteen to twenty acres), other ethnic groups, particularly the Swedes, were more apt to own their homes there. Residents who owned their homes and perhaps had a small farming operation underway had considerably more economic security than those who rented company homes. In the latter case, the families probably had gardens and kept a few animals, but without any income over a period of several months, their accounts at the company store for groceries and other necessities and the accumulation of bills for house rent, coal and water delivery, and medical dues would have been substantial. Black families like the Tates, Armstrongs, Chapmans, and Gaineses who owned land, however, remained in the area for many years.

Five months after the explosion at Mine No. 12, Buxton experienced another exodus of residents. When Consolidation opened Mine No. 18 in August 1914 and established the coal camp of Consol in Wayne Township eighteen miles southwest of Buxton, the company followed the pattern it had followed with Buxton and did not locate the new community adjacent to the mine. Rather, it located the new camp a few miles north of it. Immediately after the camp was founded, some Buxton families moved there, presumably moving into company houses. The following year several Buxton businessmen moved their operations there too, and Buxton residents continued to relocate in Consol throughout 1915 and early 1916. In January 1916 the *Bystander* noted that a Buxton resident had moved three more Buxton houses there. A former Consol resident, Mrs. Henry Brue, remembered that the camp had a large railroad operation, including a roundhouse and a depot. The community also contained one service station and several other businesses, including a hotel and a combined pool hall and movie theater. Lewis Reasby had moved there from Buxton and built the pool hall. Mrs. Brue stated that Consol had no churches but that residents sometimes held services in the schoolhouse. One former resident estimated that at its peak six hundred men worked in Mine No. 18.[5]

It is not clear what policy the company pursued in relocating employees to Consol. Apparently miners had the option of living in Buxton and commuting the entire eighteen miles daily or moving to Consol and greatly reducing the commuting time. The majority of miners apparently preferred to remain in Buxton, where they had access to a wider array of social activities and far more extensive shopping facilities.[6] Although population figures on Consol are not available, it is assumed that the camp contained between one and two hundred people.

While Buxton's population decline reflected important changes in the town's economy in 1914, company policies explain why these changes began to take place. Certainly the move of some employees from Buxton to Consol and the closing of Mine No. 12 accounted for some population loss, but during the same period Consolidation officials also made personnel changes and formulated long-term plans that in effect guaranteed Buxton's eventual demise. In April 1913 R. R. MacRae resigned as manager of the Monroe Mercantile Company, and company officials appointed E. D. Buckingham to that position. Several months later E. M. Baysoar resigned as general superintendent of Consolidation to purchase the Hoover coal mines. As a replacement the company appointed John E. Jeffries, who had formerly worked at Mine No. 18.[7] While it is not clear whether the new superintendent made any immediate changes in Consolidation's overall operation, the hiring of a new store manager did produce at least one change in the store. In November 1913 the *Bystander* editor reported on a recent visit to Buxton, noting that the new store manager had "eight or nine colored clerks in the store, not as many as Mr. MacRae." The editor also reported that for the first time in its history

Buxton had a white postmaster. He stated that E. T. Mills, a black man, had been postmaster for years and that formerly all the postal clerks had been black. With the changeover in 1913, only one black postal clerk remained. The *Bystander* editor did not relate the change in post office personnel to the company, but rather blamed the Democratic party, which he hinted was responsible for the appointments. The new postmaster was Alf Thomas, who also operated a local drugstore.[8]

In 1914, in spite of the opening of a new mine, the Consolidation Coal Company seemed to be at a low ebb. From its first appearance in Iowa in 1881 until early 1914, the company had operated at least three mines at any given time, and during some years it had four mines in operation. With the closing of Mine No. 12, Consolidation had only two mines in operation, Nos. 16 and 17. Consolidation opened Mine No. 18 later in 1914, but the following year it closed down Mines 16 and 17, with the result that for part of 1915 the company operated only one mine. In 1918, Consolidation opened Mine No. 19, and for the next seven years it continued to operate only two mines.[9]

While Consolidation's cutback to one or two mines undoubtedly reduced the amount of coal produced for the railroad, the coal mined by Consolidation did not comprise the total amount purchased by the Chicago and North Western in the state of Iowa. Since at least 1906 the railroad had also been purchasing coal from private operators. In 1906, for example, the CNW purchased coal from twenty privately operated mines in Monroe, Mahaska, and Boone counties, in 1914 it purchased coal from seven private operators in Monroe County and three in Mahaska County, and in 1919 it purchased coal from seven private operators in Monroe County.[10]

Since Iowa's state mine inspectors did not list the amount of coal taken from each mine, it is not known whether the railroad's own mines or the privately operated mines supplied the CNW with more coal. Although no company records have been located that explain the railroad's action in purchasing coal from private operators rather than enlarging Consolidation's output, several factors may have influenced its policies. First, the railroad may have discovered that it was simply cheaper to purchase coal from private operations that were already underway, thus saving the cost of developing additional mines. At the same time, the railroad may have been pursuing a "wait and see" policy. Between 1914 and 1919, Consolidation was in the process of relocating employees from Buxton to Consol and Haydock. Perhaps in 1914 the CNW was simply undecided whether it would continue to operate Consolidation's mines or gradually phase out that operation and purchase all its coal from private operators. By 1914, however, at least one thing seemed clear: Buxton's days were numbered. As Consolidation expanded its operation westward, where presumably the best coal deposits lay, it became inefficient to continue transporting miners from Buxton to mines that lay eighteen to twenty miles away. Since coal explorations took some time, Consolidation probably knew as early as 1913 or 1914 that they would eventually abandon the Buxton site and relocate its workers at a coal camp closer to Mines 18 and 19 and the proposed No. 20.

Two years after experiencing long layoffs and a sizeable loss of population, Buxton suffered another economic setback in the form of several fires. In October 1916 fire destroyed B. F. Cooper's drugstore in Coopertown. Without either a water supply or fire extinguishers, local residents were unable to put out the blaze until it had burned down more than half of Coopertown's businesses. Buildings destroyed included the Buxton Hotel (Buxton's major hotel), the Cooper Drug Store, a grocery store, the Masonic Hall and several other lodge halls, the Buxton Cigar Factory, the Granberry Tailor Shop, and an amusement hall. The fire also burned several private residences. In total, local residents estimated the loss to be around $14,000, and because of an inadequate water supply, business people had been able to insure their businesses for only one-sixth of their total value. Buxton also experienced two fires in 1917 that together destroyed a grocery store, a dance hall, a garage, a bakery, and a private residence. Although the 1917 fires were major fires, they did not have the same economic impact as the earlier fires in Coopertown.[11]

By 1916, Buxton had undergone considerable change from its original composition and appearance. In December of that year, the *Bystander* editor made his annual statewide tour of communities with substantial black populations. On his return the editor wrote:

> We next visited Buxton, Iowa, which is the king of mining camps in Iowa and perhaps the country. Until recently it has been the mecca for the colored miners. However, in the last few years, since Mr. Buxton has left, several hundreds and perhaps thousands of [black] miners have left, and many of their places taken by white miners. It seems to be the policy of the new superintendent to supplant many of the colored miners with white. However, Buxton is not what it once was, nor will it ever be.[12]

B. F. Cooper, who operated a large drugstore in Coopertown, a suburb of Buxton. Cooper was known as a strong supporter of and contributor to the Buxton Wonders. *(Courtesy of Iowa State Historical Department)*

The editor estimated that only half of the company houses were occupied at the time. He also observed that the miners were not working full-time. Because of the fires and the population decline, the editor noted that Buxton looked "like a deserted village." He reported that many black businessmen had left Buxton: "Some have gone in business in Des Moines, Cedar Rapids, Chicago, Kansas City, Milwaukee and Detroit, and still more are planning to leave soon."[13] The editor's report substantiated the changes indicated in the state census of the previous year. Although population figures are not available for 1920, interviews and other historical sources indicate that blacks constituted a minority in Buxton for the remainder of its existence.[14]

For the next two years, Buxton residents continued to enjoy many of the same activities and opportunities they had known before, particularly in regard to social activities and shopping facilities, but they could not ignore the mounting evidence that their community was slipping away. Reports in the *Bystander* provided a mixed view of the community. Each column written by the Buxton correspondent included the usual notices on Buxton's social life but also added a comment or two about the families moving away. In December 1917 the *Bystander* editor returned for another visit. He estimated that by that date Buxton's population had slipped to about 2,000. He believed that many of the miners, both black and white, had moved on to other Iowa mining camps located within a radius of twenty miles of Buxton. Still, the editor believed that Buxton remained a good place for blacks to live:

> We find that the colored people now living here are prosperous, making good wages and many are saving their earnings. The present company store employs about thirty people, of which about five or six are colored. Miss Gussie Mardis has charge of the dry goods department. . . . Miss Smith is the cashier.[15]

In 1918 the exodus from Buxton appeared to be in full swing. By that year all the mines close to Buxton, including Mines 12 through 17, had been closed down. Moreover, Consolidation was expanding its operation at Consol and had opened Mine No. 19 and its accompanying camp, known as Haydock, the previous year. Haydock was located a few miles southwest of Consol, and when Consolidation built Haydock it also constructed another camp, known as Bucknell, adjacent to the site. Informants recalled that Bucknell was much smaller than Haydock and that Consolidation moved its main office there. The local high school was also located in Bucknell, though the company constructed or relocated its company houses in Haydock. In 1918 many Buxton residents moved to Consol or Haydock, while others departed for major cities in and outside of Iowa. In March the *Bystander* reported that Dr. C. G. Robinson, "one of Buxton's best physicians, leaves this week for Chicago where he intends to make his future home." Dr. Robinson had been in private practice in Buxton for several years. The same article noted that many members of St. John's

African Methodist Episcopal Church had left Buxton in the last three months. A year later Dr. Linford Willis and his family moved from Buxton to Des Moines, where the dentist resumed his practice.[16]

In August 1919, Charles S. Nichols, an Iowa State College professor of sanitary engineering, visited Buxton and reported that the camp's population did not exceed four hundred. He noted that there were several hundred houses unoccupied, "many of them in such condition as to be uninhabitable. Many are being torn down and salvaged."[17] He described the camp as generally in a state of decay:

> It is reported that in years past the Camp was regularly organized, with local authorities who maintained general sanitary order in the Camp. In its present condition, such organization, if any still exists, has become lax in its operation. Yards and streets are littered and grown to weeds; no system of garbage or rubbish removal is in vogue; and the residents are of such disposition that most of them will not attend to such matters themselves. As a consequence, large quantities of litter and rubbish are evident, the privy vaults are full, and garbage has been dumped in back yards and in alleys. The inspection was made after continued dry, hot weather, and the odors were extremely unpleasant, most particularly in the sections more thickly settled.[18]

Nichols added that the water supply was "not particularly satisfactory," because the company pumped water from a wooden tank into individual cisterns. Because the cisterns were not filled until they were nearly empty, they became stale and "as they are seldom, if ever, cleaned, there must be an accumulation of materials in them." Nichols also noted that the pond located in the southwest part of the camp had been condemned several years earlier both as a water supply and as a source of ice for domestic use.[19]

According to Reuben Gaines, Jr., and Clayborne Carter, the company continued relocating Buxton residents throughout 1919, 1920, and 1921. By December 1922 all the miners had been moved from Buxton. Company officials had moved most of the Buxton houses to Haydock and had sold the remainder. Reuben Gaines, Jr., recalled that these houses had been sold for $100 for a five-room house and $125 for a six-room house. Some Buxton business structures were purchased by neighboring residents and moved to new sites, while others were torn down. The company did continue the miners' train until 1923, however, as Archie Harris rode the last miners' train from Buxton to the outlying camps of Consol and Haydock in September 1923. Harris recounted that by that time all the buildings in Buxton had been torn down or moved away.[20]

Haydock, like Buxton before it, expanded rapidly. Local residents estimate that before Haydock was founded in 1917 only four families lived in the area. But within a year the boom was on, and Haydock eventually rivaled Buxton in size. Former residents estimate that Haydock soon contained approximately 900 houses and had close to 6,000 people. Housing in Haydock was mostly company owned and had two origins. Consolidation built some new homes, although the number is not known. Apparently company officials believed that the camp would survive for some time, as they constructed five-room homes on poured concrete foundations. The miners' homes contained neither electricity nor indoor plumbing, however. Pictures of the camp indicate that homes were built along the same design as Buxton company houses. Consolidation also built homes for their management personnel that included basements, furnaces, and sidewalks.[21]

Most company houses in Haydock, however, had been moved from Buxton. Mike Onder recalled that in moving the houses the crews "just cut [the houses] in sections . . . and put them on a flat car, . . . moved them out and put them back together in sections." Earl Smith also recalled that the company had moved many houses from Buxton to Haydock: "They'd move them out there with mules on hayracks, flatbed wagons, and then later they started to tear them down in sections . . . loading them on the railroad and hauling them out. They moved a lot of houses out there from Buxton, and that's where most of the houses come from [in Haydock]." Archie Harris recounted that Consolidation hired crews of seven men each to dismantle the houses and each crew dismantled two houses a day. "They tore the houses down," he said, "and pulled the nails out of it and stacked the lumber up there and that was one day's work." He recalled that each man was paid five dollars a day.[22]

Consolidation invested only a modest amount in public improvements in Haydock. Company workmen planted trees along the streets and Consolidation officials talked of paving the streets, but that did not materialize. Lester Beaman remembered that the town had dirt roads and no streetlights. For water, the company built a series of cisterns scattered throughout the town. Housewives came to the cisterns to collect their daily water supply.[23]

Haydock soon attracted many business people, including some who had been in business in Buxton. At its peak the town had two hotels, a drugstore, a company store, a lumberyard, a soft-drink factory, two theaters, a bakery, three garages, and seven pool halls.

Haydock also had seven churches, several grade schools, and a three-story high school. Earl Smith recalled that the high school had grades nine, ten, and eleven. Mildred Hight Covey, whose father served as the Methodist minister in Haydock, attended high school there. She recalled that the high school had both black and white students and that the school had an active basketball program for both male and female students. Haydock also had at least one restaurant, owned and operated by C. W. Armstrong, a son of Hobe Armstrong. Armstrong also operated a general store there. Lester Beaman related that the Armstrong store was large and that "they had everything." When the London family relocated in Haydock, W. H. London operated a grocery store and his wife, Minnie, taught school. The Langlois sisters also relocated in Haydock; they operated a dance hall and opened another movie theater. At least one company physician, Dr. W. S. Chester, moved his practice to Haydock.[24]

Many individuals also moved buildings from Buxton to Haydock. Jake Wilson purchased several company houses, moved them to Haydock, and rebuilt them into a hotel. Hobe Armstrong bought one of the black churches in Buxton, moved it to Haydock, and remodeled it into a store. At the same time the congregation of the Mt. Zion Baptist Church moved their church building to Haydock and continued the church there. One of the major businessmen in Haydock was Reuben Gaines, Jr. When workmen tore down the Monroe Mercantile Company in Buxton, Gaines purchased some of the lumber and used it to build several buildings in Haydock. In his own account of Buxton, Gaines wrote that he also purchased a large grocery store at Miami and six houses in Buxton, all of which he relocated in Haydock. He also constructed a building in Haydock that measured 30 feet by 120 feet, with two stories and a half-basement. He related that "the first floor was used as a pool hall, barber shop, confectionaries and other purposes. The second floor was the dance hall and where eight lodges met and the miners' Union from 18 mine." Gaines added that "the next investments were in buying a number of lots and building a home."[25]

Although it is not known precisely how many families moved from Buxton to Haydock, interviews indicate that the number was substantial. Hucey Hart estimated that approximately three-fourths of the population of Buxton relocated in Haydock. There is no indication that Consolidation did any further recruiting, either in Iowa or in the southern states.

For some families, life in Haydock was much the same as life in Buxton. Lester Beaman related that while living in Buxton his father had lost his eyesight as a result of a mining accident. Beaman's father was then unable to work and his mother took in washing to support the family. After relocating in Haydock, Mrs. Beaman continued to take in washing. Susie Robinson and her children and parents also moved to Haydock. Mrs. Robinson remembered that "Haydock was another great big fine place, but it didn't last long." Many families who moved from Buxton to Haydock remained there until the camp closed down in 1927.[26] Given the normally high rate of geographic mobility among coal miners, some mining families probably also moved from surrounding camps once Haydock got underway.

While Haydock residents did not enjoy the social opportunities that had existed in Buxton, residents did have some social outlets. Hucey Hart remembered that the building erected by Reuben Gaines, Jr., was a popular place. "That was our recreation. We miners would go up there, you know, and we'd play cards . . . and pool and stuff like that. But it wasn't nothing like Old Buxton. See, you didn't have too many places to go like you did in Buxton." Nellie King related: "We'd give dances in Haydock. We'd have the Marigold Strings—that was the Bakers' boys. We knew them from Buxton. And they'd play for us for forty-five dollars a night. We'd have to guarantee 'em that. If they played till 1:00 A.M. we'd have to pay them sixty dollars." Mrs. King remembered that food and different soft drinks were sold at the dances. Reuben Gaines, Jr., provided the facilities for lodges, many of which had relocated their organizations from Buxton.[27]

In general, harmonious race relations also carried over from Buxton to Haydock. Hucey Hart remembered that blacks and whites continued to get along well together. Nellie King went to high school in Haydock and all her teachers were black. She stated that all the schools in Haydock were integrated. Lester Beaman, when asked about race relations in Haydock, responded: "Well I think [blacks and whites] got along pretty good. . . . I know they did in Haydock, I was a little older in Haydock. We got along real good, . . . never did have no trouble." Beaman did not remember that there were any all-white neighborhoods in Haydock, but rather that all families lived in an integrated setting.[28]

In 1923, Consolidation moved its office from Buxton to Bucknell, but it was to remain there for only two years. On November 23, 1925, Consolidation's stockholders held a special meeting in Bucknell to consider the sale of Consolidation to the Superior Coal Company, an Illinois corporation. The stockholders voted in favor of the sale with the clear intent that the action would bring about the "immediate and complete dissolution of Consolidation Coal Company." The sale included all of Consolidation's property in Iowa, including land, coal rights, coal deposits, buildings, mines, and railroad tracks, and all cash on hand. They voted that all money from the sale, as well as the remaining assets of Consolidation, be divided up among the stockholders. The sale price was $747,629.30.[29]

The reasons behind the sale of Consolidation are not entirely clear. According to company records, Consolidation was concerned about higher tax assessments levied by Monroe County officials in 1925. In fact, Monroe County officials had assessed the company tax penalties of more than $10,000 for 1923 and 1924, and a settlement had been reached between Consolidation and Monroe County officials whereby Consolidation paid slightly more than $5,000 in fines. In 1925, Consolidation officials expressed concern that the officials in Wayne Township, where Haydock and Bucknell were located, had fixed the valuation of the Consolidation capital stock for taxation purposes at $30,000 for that year. This meant that Consolidation would pay property taxes of $1,600. This action apparently angered company officials, as they believed that other mining companies were paying lower taxes and that local tax officials were therefore discriminating against Consolidation. In his memoirs of Buxton, Reuben Gaines, Jr., explained that Consolidation had some difficulty with the tax officials in Monroe County over the closing of Mine No. 12, but he does not state the exact nature of the problem. Gaines also relates that at one point Monroe County officials "drove out to Wayne Township and closed [the Consolidation mine] because the Coal Company had failed to pay a $30,000 tax on time." When the controversy over taxes arose in 1925, company officials considered several options, including transferral of Consolidation's place of business outside Monroe County, and finally settled on the decision to sell the property to the Superior Coal Company.[30]

Although Consolidation sold out in 1925, it seems that the relationship between Superior and the Chicago and North Western Railroad was the same as that between Consolidation and the railroad. In 1925 the state mine inspector for District Two (which included Consolidation's mines in Monroe County) listed the Superior Coal Company as having two mines, Nos. 18 and 19, operating in Monroe

County. Consolidation's general superintendent, John Day, was retained by Superior, and the Chicago and North Western Railroad was listed in both 1924 and 1925 as the firm purchasing and shipping the coal for both Consolidation and Superior, indicating that the Superior Coal Company, like Consolidation, was a subsidiary of the CNW. Also supporting this conclusion is the fact that not a single person interviewed referred to any sale of Consolidation in 1925. In fact, many interviewees talked about Haydock after 1925, particularly the closing of the mine in 1927, but all continued to refer to the company as Consolidation. If the relationship with the Chicago and North Western had been severed in 1925, it is likely that the new company would have implemented some new business and management practices. However, this does not seem to have been the case. The Superior Coal Company remained in Iowa for two years, operating Mines 18 and 19, maintaining its headquarters in Bucknell, and housing its workers in Haydock.[31] In effect, it seems that Superior replaced Consolidation as the CNW's coal mining subsidary in Iowa from 1925 to 1927.

Although there is no evidence that Superior made any significant changes in its treatment of employees or in manager-employee relations, employees did experience fewer working days because of external conditions. In 1925 a state mine inspector noted that because of the extreme competition from eastern coal, the operators of many larger Iowa mines (shipping mines) had adopted the policy of shutting down about April 1 rather than operating throughout the summer. These mines then reopened in August to prepare for the fall and winter trade. This marked a significant change for shipping mines, which had previously worked year-round, though summer work may have been somewhat less than winter work.[32]

As early as 1919, the national coal industry had begun to suffer a gradual decline in the demand for coal. Increasingly, coal operators sold less coal as more and more Americans used natural gas, electricity, and petroleum for heating and industrial purposes. The greatest problem facing the national coal industry, however, was overdevelopment. Iowa state mine inspector R. T. Rhys succinctly stated the problem in 1924:

> There is not the slightest doubt but what the coal mining industry [nationally] in the last seven years had been overdeveloped until there is at the present time an excess of mines and miners to produce the coal requirements of the country and it is inevitable that a deflation in mines and miners is an economic necessity.[33]

Problems within the Iowa coal industry itself also made the 1920s difficult years for both Iowa operators and miners. Iowa coal operators had been combating a negative image of Iowa coal for over a decade. Increasingly after World War I, Iowans had been purchasing coal produced in eastern states, particularly Kentucky, explaining that out-of-state coal burned cleaner and produced a hotter flame. In 1924, Iowa mine inspector R. T. Rhys wrote in his annual report: "It is to be regretted that the coal mining industry of our state is not receiving from some of our people the consideration and the patronage it deserves. The long and continued depression of the industry has brought many of the coal operators face to face with financial ruin, and a large number of the miners and their families to the verge of want." Rhys added that in the early 1920s residents of Iowa consumed over 14,000,000 tons of coal but that the total production in Iowa mines for any given year was far below that amount. In 1923, for example, Iowa coal mines produced only 6,120,332 tons. That meant that in 1923 over 8,300,000 tons of coal were shipped into Iowa. Rhys estimated that of the amount shipped in from other states (principally Illinois, Indiana, Kentucky, West Virginia, and Pennsylvania), about 50 percent was purchased by the railroads and the other 50 percent by domestic and industrial users. Rhys concluded that in light of these figures, residents of Iowa (not including railroads) were purchasing almost twice as much coal from out-of-state producers as they were from Iowa producers.[34]

To Rhys and many others in the Iowa coal industry, the most frustrating aspect was that while Iowans purchased thousands of dollars worth of coal from eastern producers, many Iowa coal operators and miners were sitting idle. In his 1924 report, Rhys lamented:

> What is more unreasonable and unfair than to see people, and especially businessmen, who are living in what may be termed mining communities, buying their coal from Kentucky or some other Eastern state, when the mines of that neighborhood are idle because of the lack of demand for coal, and when likely as not the operators and miners of that locality are buying their groceries and clothing from them?[35]

It was this combination of problems in the state and national coal industry that led to the downfall of Haydock. During the 1920s, as miners began to work fewer days, some began to drop their union affiliation, thus placing pressure on UMW officials to negotiate the best possible contract to keep the remaining miners in the union fold. Because of the inability of the UMW and the coal operators to agree on a contract in 1927, miners all over the country, including those in Iowa's District 13, walked off the job on April 1.[36]

Interviews with former Haydock residents indicate that because of this strike Superior shut down its operations in Iowa. Hucey Hart remembered that event:

Haydock blew up in 1927. They went on strike and [company officials] told them if they went out on strike they would stay out. They'd close the mine down. Well, [the miners] didn't believe it, so they went out. We was out about a year and a half and they closed the mine down. Brought all the mules on top, took dynamite and blowed the top down. They left all the rest of the coal and the motors and things down inside the mine.[37]

Oliver Burkett also remembered of the 1927 strike that the miners "never did go back to work. Papa fooled around there a year. He came to Waterloo in 1928 and got a job at Rath, and we stayed down there another year before we moved here, before he sent and got us."[38] In his memoirs, Reuben Gaines, Jr., captured the feeling of dismay and frustration felt by the local miners:

[The closing of the mines] left many Iowa miners in unusual circumstances. Many of the miners waited nearly two years . . . for the mines to reopen, they were not working and they were hungry so they ventured out at nite and stole hogs and anything else that was edible. During the waiting period when the miners thought the mines would reopen again there were large size groups that would argue and debate the issue day by day in my Pool Hall. I always said that the mines would never open from the very beginning so they were never interested in my advice but I gave it freely for I said I was going to leave before I got too thin and you [the miners] should [too].[39]

Gaines, like the others, perceived the mine closing as a direct result of the strike. He concluded: "In other words it was the last straw that had broken the camel's back or the last drop of water that over ran the cup."[40]

The closing of Mines 18 and 19 had other far-reaching effects. Former Haydock residents recalled that when the strike started the company was in the process of sinking the shaft for No. 20, but, as Carl Kietzman recalled, the company never opened it: "They laid the track up to it. They had labor problems and [the company] took their business down to southern Illinois, around Gillespie and Bunnell." Mike Onder also recounted that the company had started to sink No. 20, but after they got it about half sunk they simply stopped work. They then pulled up the track and, according to Onder, "that ended

An abandoned house at the Haydock site, 1963. The structure may have been a company house changed slightly, with a room added on the right side. (*Courtesy of Iowa Mines and Minerals Department, Des Moines*)

that." Reuben Gaines, Jr., remembered that the company had had plans not only for No. 20 but also for a No. 21, 22, and 23.[41]

With the departure of the Superior Coal Company, Haydock soon became a ghost town. The company sold off its houses at fifty dollars each and "real estate men bought them by the dozens, knocked them down in sections and hauled them away on flat cars." The usual practice was to relocate the homes in nearby communities like Chariton and Albia, where they would be resold. Some businessmen remained in Haydock, hoping that the town would retain enough population to maintain a few businesses. That hope was not realized, however, as local residents estimated that within six years the population had dropped from 6,000 to less than 100. The last surviving institution was the high school, which remained open until 1935. Twenty years after its founding, little remained of Haydock except scattered house foundations and some abandoned pieces of furniture. Like Muchy and Buxton before it, Haydock had achieved some prominence as an Iowa coal community, but the deep-rooted economic problems of the coal industry meant that Haydock would not survive for long.[42]

William and Minnie London were two of the many residents who left Haydock in 1927. The Londons, like many other families, had spent over thirty years associated with the Consolidation Coal Company. They had both arrived in Muchakinock in the early 1890s. Minnie Robinson had arrived in 1891 as a single woman. She took a teaching position and a short time later married a local businessman, William London (better known as W. H.). The Londons lived in Muchy until 1901, when they moved to Buxton. There they both continued their occupations. Mr. London worked for a time at the company store but eventually opened his own store; Mrs. London taught school and later served as both teacher and principal. As a teacher, Mrs. London occupied a position of esteem within all three communities, most notably in Buxton. During their years there, the Londons raised two children, Hubert and Vaeletta, and many *Bystander* articles reported on the dinners and social affairs hosted by the London family. Unlike most Buxton residents, the Londons sent their children to college. Both children attended the University of Iowa, where Hubert received an M.D. and Vaeletta majored in English. Around 1919, the Londons moved to Haydock, where once again they each pursued their lifetime work: W. H. opened a store and Minnie taught school. When Haydock closed down in 1927, the Londons moved to Des Moines, where they joined their son. In 1939, following her husband's and son's deaths Mrs. London joined her daughter in Waterloo. She remained there until her death in 1955.[43]

The experience of the London family serves as a good case study for the geographic mobility of the residents of Muchakinock, Buxton, and Haydock. While not all residents of Muchy lived at a later time in both Buxton and Haydock, many did so. Like the London family, many residents spent almost half a lifetime associated in one way or another with the Consolidation Coal Company. Many Buxton residents chose not to move on to Haydock, however, and these people relocated in all parts of the United States. While most community studies are concerned only with a community's population during the time the study is being made, in this instance it seems worthwhile to examine the movement of some residents after they left Buxton, and the interviews provide an opportunity to trace many families through three or four decades of their existence. In this way, the social and economic experiences of some of the families in Buxton can be compared with their later experiences in communities like Des Moines and Waterloo. While Buxton is the focus of this study, a long-term perspective acknowledges that these people's lives did not come to an end in 1920 or 1927. In effect, an examination of these families after they left Buxton provides the opportunity to carry the Buxton story to its logical conclusion.[44] While many black families, like the Londons, had lived in considerable comfort and had enjoyed many social amenities during their days in Muchy, Buxton, and Haydock, the move away from the coal camps brought some into contact with racial discrimination. It is not known what experiences W. H. and Minnie London may have had, but their daughter, Vaeletta, experienced a new world after moving to Waterloo. Following her graduation from the University of Iowa in 1917, Vaeletta married Milton Fields. Fields had completed a law degree at Iowa and, following his military service in World War I, the young couple moved to Waterloo. The Fields were part of a large black community there, many of whom had arrived in Waterloo in 1911 after migrating from the South. In that year the Illinois Central Railroad had hired black men as strikebreakers when white workers had walked out to protest wage cuts. The blacks quickly built shanty-type homes close to the railroad tracks and lived there for many years. Some community services were denied them, but gradually black families began to recruit their own people to provide these services.

Milton Fields had come to Waterloo to practice law, and he had little difficulty in attracting black clients. He soon discovered, however, that they frequently had little money with which to pay his fees. Because of this, Vaeletta realized that it would also be necessary for her to work. Although she was a college graduate with a degree in English, she knew from experience that the Waterloo public schools would not hire her as a teacher. (The Waterloo public schools did not hire its first black teacher until 1952.) She went to work as an elevator operator and later as a cleaning woman in the women's rest room of the local telephone company. Later she worked as an assistant to the YWCA director in Waterloo.[45]

Black coal miners who wished to continue as miners after leaving Buxton also experienced some racial discrimination. Often they discovered that they were not welcome in other coal mining communi-

Milton Fields and Vaeletta London during their courting days. The couple would later reside in Waterloo. (*Courtesy of Vaeletta Fields*)

ties. Some Iowa coalfields remained all-white into the 1920s, a situation that apparently reflected the attitudes of both the local residents and the coal operators. Residents in Seymour, for example, stated that no blacks resided in Seymour during the years that coal mining was an active industry there, and a retired Buxton miner stated that no blacks were allowed in Bussey. As discussed earlier, in 1909 Ogden underwent severe labor strife when 160 black miners from Buxton and Oralabor arrived to take jobs in the all-white coalfields there. Local white residents expressed extreme hostility toward them, and within a year the black miners had returned to their former homes. In some coal camps, such as Zookspur (located several miles south of Madrid), black families were congregated together in one area of the camp.[46]

Some Buxton residents who left in the teens relocated in major industrial cities in the Northeast. Two of Gertrude Stokes's brothers settled in Detroit; one worked for the sanitation department and the other worked in the automotive industry. Both brothers settled in Detroit permanently. Gertrude Stokes and Hucey Hart remembered that many people went to Milwaukee after leaving Buxton. Others settled in Washington, D.C., and Chicago, and some traveled west to California.[47] Many black families who left Buxton in the late teens resettled in other parts of Iowa, particularly in Des Moines, Waterloo, and Cedar Rapids. In those cities black men went to work in packing plants and auto assembly plants or worked as clerks, janitors, and day laborers. Some opened businesses, particularly food establishments and barber shops. Lara Tolson Wardelin was one of the many former Buxton residents who resettled in Des Moines. Lara and her family left Buxton in 1912 to live in Whiteburgh, a nearby coal camp. They came to Des Moines in 1916. Lara's father went to work as a cook at a private club, an occupation that he had pursued before becoming a coal miner. He worked for several years in Des Moines, but when his health began to fail, both Lara, who was in eighth grade, and her brother had to drop out of school and go to work. Lara worked as a domestic both before and after her marriage at age nineteen. Her work required that she remain in her employer's home until she had washed the supper dishes; she then went home by streetcar or bus. In the meantime, Lara's husband had fed their four children at home. In later years Lara worked in a variety of occupations, including work as a tailor, laundress, and cook.[48]

For some families, assistance from family members was crucial in successfully relocating after leaving Buxton. While her family still resided in Buxton, for example, Odessa Booker's father died, leaving her mother with many small children. Her mother's brother lived in Des Moines, and he decided that it would be better for the Booker children if they all moved there. Mrs. Booker remained in Buxton for about a year after her husband's death and then, with the help of her brother, relocated in Des Moines. Odessa remembered that her mother decided to move because "we didn't know anybody else down there. We didn't have any more family down there." The Booker family moved to Des Moines, where their relative helped them obtain a house. Odessa's mother lived in the same house until she died many years later.[49]

Kinship ties also remained strong among other families and sometimes determined that all family members would relocate together. Alex and Agnes Erickson lived with their parents in Buxton until the camp closed down. Then the parents, Alex and Agnes (both of whom were single), and a married sister and her husband all decided to relocate in Pershing, a coal camp approximately fifteen miles northwest of Buxton. When it came time to move, Alex and his father dismantled their house in East Swede Town, moved it by truck, and set it up in the Pershing camp. The Erickson's married daughter, Dena, and her husband also moved a house from Buxton and relocated it on an adjacent lot.[50]

For some black families who had experienced economic hardship in Buxton, that hardship continued in other places. Mattie Murray related that when her father, Tandy Bradley, died in Buxton, her mother was left with fourteen children, all at home. Mrs. Bradley then moved her family to a coal camp near Knoxville where she could rent a house for five dollars a month. Mrs. Bradley obtained work as a cook in a Knoxville hotel, and every day, summer and winter, she walked the three and a half miles to Knoxville. Mrs. Murray remembered: "I seen her walk on many a day and when she got home her clothes were froze up to here in the wintertime, and she'd have baskets on her arm." She added, "My mother had to work like night and day taking care of all of us, and she worked and she didn't get but ten dollars a week cookin' at a hotel right down here in Knoxville." She remembered, however, that some people were sympathetic and helpful to her mother. In the evening when Mrs. Bradley left the hotel, the hotel manager would say, "Mrs. Bradley, everything that's left over, you put it in a basket and take it to your children."[51]

For other black families, hardship came in different forms. Marjorie Brown left Buxton at age twelve and moved with her grandparents to Cedar Rapids. Marjorie's uncle had heard about work there in the starch plant, and he and another relative traveled to Cedar Rapids to check it out. One went to work at the starch plant and the other at Quaker Oats. When they wrote back and said there was work, Marjorie's grandparents and a number of other Buxton people headed for Cedar Rapids. Having come from a prosperous home in Buxton, where both the Brown parents and grandparents had been part of the Buxton elite, life in Cedar Rapids was difficult to accept. Mrs. Brown stated that when she arrived there as a twelve-year-old, she was accustomed to taking part in all school activities, particularly music. "So when I went to school in Cedar Rapids," she recalled, "they were gathering up the children that could do things, you know, that could play or could sing for the little choruses and all. I said I could play and I could sing and that's the first time in my life that I have ever been looked through rather than at. Yes, nobody told me I couldn't take part. I was just looked through. I didn't exist and I wasn't there." Mrs. Brown stated that up to that point it had always been: "Marjorie can do it. Marjorie can play. Marjorie can sing. . . . And then all at once, with no warning, I no longer existed."[52] Following her marriage in 1922 at age sixteen, Mrs. Brown lived in several communities, but in 1934 she and her husband came back to Cedar Rapids. The depression of the 1930s proved to be the most difficult time for the Browns. Mrs. Brown's husband, although a college graduate, went without work for two years. She remembered the time as:

> one of the bitterest periods of my life. . . . I'm one of the people that sat in their house . . . or stood and watched the house full of furniture be taken to be reclaimed. If there hadn't been a stairway, I'd sat on the floor. I have worked seven days a week doing housework, and that includes washing, ironing, cooking—and believe me, we did do windows—for five dollars a week. Because it was two years, almost two years, that Mr. Brown didn't have any work. I don't mean that he didn't work a day or two. He had *no* work.[53]

Although economic conditions improved somewhat for the couple, Mrs. Brown continued to work for many years. In 1951 she moved to Waterloo.[54]

For at least a small number of people, life after Buxton proved to be at least as pleasant as it had been before. Marian Carter, Dr. E. A. Carter's daughter, described her father's life and her own after the family moved from Buxton in 1920. The family moved to Detroit, where Dr. Carter continued to practice medicine. He served on many civic committees and belonged to many groups there. He was on the first Mayor's committee on Interracial Affairs, and he belonged to the NAACP, which he addressed frequently. In addition, he served for many years as the examining physician for the Supreme Liberty Life Insurance Company of Chicago. Dr. Carter died in Detroit at the age of seventy-five. Marian worked in Detroit as a public school librarian and counselor for many years.[55]

Two of Buxton's black professionals resettled in Des Moines. Dr. Linford Willis moved his dental practice there in the late teens. George Woodson moved to Des Moines in the early 1920s and contin-

Dr. Edward A. Carter after his move to Detroit where he continued his practice of medicine. (*Courtesy of Iowa State Historical Department*)

ued his law practice. Woodson remained active in state and national bar associations, but there is no evidence that he continued the active life in the Republican party that he had pursued in Buxton. He died in Des Moines in the mid-1930s.[56]

Some individuals remained within the vicinity of Buxton. Reuben Gaines, Jr., after losing heavily on his investments in Haydock, moved to a rural area north of Albia, where for many years he maintained a successful turkey farm. Many other individuals remained even closer to the Buxton site. After the Buxton mines closed down, Archie Harris became a full-time farmer, relocating on an area that was once a part of Buxton proper. The Emmanuel Blomgren family also settled close to the original site.[57]

In the late teens and twenties, then, the populations of Buxton and Haydock, like that of many other coal camps, scattered throughout Iowa and the nation. In general, white families tended to remain in Iowa, often moving to other camps or turning to full-time agricultural pursuits, while black families demonstrated a greater geographic mobility. For blacks, Muchakinock, Buxton, and Haydock had been stopping places on their way from the South to the North. In essence, these three communities had acted as collecting points for large numbers of blacks who later radiated out to all parts of the northern United States. For most black families, life after Buxton and Haydock was less prosperous and less fulfilling than the life they had known in Monroe County. Many black former residents revealed that they experienced their first racial prejudice after they left the employment of the Consolidation and Superior coal companies, though for at least a few the good life continued. As residents migrated out of Monroe County to relocate in other localities, their migration signaled more than the end of two mining communities; it signaled the end of a prosperous Iowa coal mining industry.

8

A PERSPECTIVE

IN 1963, Hubert Olin, who was then engaged in research on the history of Iowa's coal mining industry, spoke before a reunion of Buxton residents. Olin asked his audience: "What was the quality of that old camp that sets its traditions so much above those of a hundred other ghost towns?"[1] Olin's question goes to the heart of the Buxton story. In the first two decades of the twentieth century, the community of Buxton exhibited social and economic characteristics that were highly unusual both for Iowa and the nation. Set down in the midst of a prosperous agricultural state, Buxton became the center of a large industrial enterprise that provided hundreds of mining jobs for blacks and other ethnic groups. This development resulted in an ethnic pluralism unknown in other parts of Iowa. While dozens of coal camps came into existence and quickly faded away, Buxton attracted national attention because of its hospitable environment for blacks. Heralded as one of the great mining camps in the country, Buxton was widely known as a model mining community where blacks and whites lived side by side in racial harmony.

In order to understand the Buxton experience more fully, it is necessary to place the community in a wider context. In effect, Buxton was only one of three coal communities supported by the Consolidation Coal Company. The first, Muchakinock, was not established by Consolidation but came under its control in 1881 when the company purchased coal interests in the Muchy area. Rather than developing as a planned community, Muchakinock grew in haphazard fashion. The practices that Consolidation established in Muchakinock, however, were later carried over to Buxton, greatly affecting the way the work force and their families lived. In Muchy, Consolidation introduced the practice of hiring large numbers of black workers, and it apparently perceived little, if any, difference between the work performance of blacks and whites, because it not only continued to hire blacks there, but it transferred that practice to Buxton and later to Haydock. Muchy attracted many black families whose offspring later worked both at Buxton and Haydock. Many of these families attained some economic and social prominence. In fact, Muchakinock attracted many individuals who provided leadership for the black community there and later in Buxton.

Buxton, the second stage in Consolidation's operation, represented the culmination of policies devised in Muchakinock. In effect, the Buxton phase represented the great experiment. With the relocation of workers from Muchy and the recruitment of additional southern blacks, Consolidation apparently intended Buxton to be a predominantly black community. The company made a sizeable commitment to its black employees with the construction of two YMCAs and carefully planned the physical layout of the community. As a result, Buxton differed significantly from most coal camps in regard to its general appearance and design.

Buxton flourished between 1900 and 1914, attracting widespread attention because of its size, community institutions, and racial policies. By 1914, however, the company had apparently abandoned its concept of a predominantly black community, for reasons that are not entirely clear. White miners began to replace blacks and fewer blacks were hired in the company store. Yet, from 1914 until the company abandoned the community in 1923, Consolidation continued to support the YMCAs and to treat blacks and whites equally in employment and housing.

Haydock, which existed from 1918 until 1927, represented the third stage in Consolidation's operation. Like Muchy before it, the company moved employees from Buxton to Haydock, which meant that the population of the latter community contained many black residents. Although the company believed that Haydock would be of considerable duration, they did not build a YMCA there. Apparently by the late teens the company still perceived blacks to be good workers, but they did not see the need to create special community institutions for them. Although it was larger than most, Haydock was essentially just another company town in southern Iowa. The company apparently continued its policy of equal treatment for blacks and whites, as many black families remained there until the community disbanded.

While the Consolidation Coal Company (and later Superior)

took care of general planning and handled the day-to-day affairs in all three communities, its parent company, the Chicago and North Western Railroad, served as the sovereign power. As a subsidiary of the railroad, Consolidation existed at its pleasure. As long as the railroad needed Iowa coal, the Chicago and North Western maintained its Iowa properties. But in the decade of the twenties, the railroad, through its subsidiary companies, made several decisions that resulted in the closing of its Iowa properties. During that decade, the CNW, along with other major Iowa railroads, began purchasing larger and larger amounts of coal from out-of-state producers, and thus its dependence on its Iowa holdings gradually decreased. As early as 1915, in fact, Consolidation apparently had some doubt about continuing its coal operations in Monroe County, as during part of that year it operated only one mine there. During the following decade, Consolidation operated two mines in Monroe County but did not expand to its previous level. In the 1920s, the company also experienced labor difficulties at its Iowa mines. In 1927 employees at Haydock honored the order of the United Mine Workers of America and went out on strike. As a result of that action, railroad officials, through their apparent subsidiary, the Superior Coal Company, closed down Mines 18 and 19 and stopped work on No. 20. With that move, the CNW effectively closed down its Iowa mining operations.[2]

The shutdown of the coal mines at Haydock in 1927 helps to place the Buxton experience in its proper perspective. Although questions might be raised as to the altruistic nature of either the mining subsidiaries or the railroad officials, the establishment and maintenance of its coal camps was first and foremost a business proposition. Nothing indicates that more clearly than the move in 1927. To Consolidation, Superior, and the Chicago and North Western, the coal camps of Muchakinock, Buxton, and Haydock were business operations and nothing more. When these communities ceased to fulfill their economic purpose, they were quickly abandoned.

During the time that the communities existed, however, they proved to be of considerable worth, not only to the railroad but also to officials of Consolidation. The company provided members of the Buxton family and other company officials with an affluent life-style and also allowed them to extend at least some degree of that affluence to a host of relatives. While the salary of the general superintendent is not known, it is known that the housing provided for his family was almost palatial when compared to the miners' housing. The Buxton

Consolidation Coal Company's Mine No. 19 at Bucknell. This was the last mine opened and operated by Consolidation. (*Courtesy of Iowa Mines and Minerals Department, Des Moines*)

family held stock in the Buxton Savings Bank and probably also had other business interests in the town. The positions of John Buxton and his son, Ben, as general superintendents of Consolidation spanned a period of almost thirty years. These positions and business interests allowed the elder Buxton to maintain a summer home in the Green Mountains of Vermont and his son to maintain a home in Florida. A later superintendent, E. M. Baysoar, also apparently did well financially, as in 1913 he left Consolidation to purchase the nearby Hoover mines and become a coal operator in his own right.[3]

For the Buxton family in particular, employment with Consolidation allowed them the opportunity to extend their good fortune to many relatives. In Muchakinock, Dr. Henderson, a company doctor with Consolidation, was a brother-in-law of John Buxton. W. H. Wells, the manager of the company store in Muchy and later in Buxton, was a son-in-law of the elder Buxton. C. N. Paris, who worked for the company for many years as a weighmaster, was a nephew of John Buxton. Cora Paris King, daughter of C. N. Paris, worked as a bookkeeper and saleswoman in the ladies' ready-to-wear department of the company store. Some relationships carried over to Haydock. Cora King's husband, Fred, worked in Haydock, where he managed the company store, bank, and post office.[4]

Of all the factors related to Buxton's success as a prosperous, racially integrated community, the time period itself must be considered paramount. The turn of the century was a propitious time for the community to begin. In 1900 the Iowa coal industry was in an expansionary period and times were prosperous. The Iowa industry relied heavily on railroads as the major consumers of coal, and most major Iowa roads purchased Iowa coal. In fact, it was the expansion of the railroads during the last two decades of the nineteenth century that led to the initial expansion of the Iowa coal industry. As long as railroads continued to purchase large amounts of Iowa coal, the Iowa coal industry remained healthy.

The matter of trade unions should also be considered in regard to the period in which Buxton originated. The rise of the United Mine Workers of America in the 1890s paved the way for better working conditions at Buxton. In 1900, Buxton miners voted to join the UMW, as did most other coal camps in the state. With strong backing from Iowa's coal miners, UMW officials successfully negotiated with coal operators for a series of changes in the coal industry that resulted in better working conditions for Iowa's miners and greater job and wage security. Specific changes included the hiring of separate shot firers and shot examiners (these usually had been a combined position), which resulted in safer working conditions, and the acceptance of the joint wage agreement. The latter guaranteed workers equal wages for equal work and also frequent wage increases. The wage agreement also greatly reduced the number of wildcat strikes, which had proved disruptive and costly to operators and miners alike. The employees of the Consolidation Coal Company in both Buxton and Haydock enjoyed better working conditions and higher wages because of the presence of the United Mine Workers in Iowa after 1900.

The prosperous times in the Iowa coal industry, moreover, resulted in greater union stability. In times of economic distress, miners—like many other occupational groups—tended to drop their union affiliation. When good times returned, labor organizers again faced the task of organizing the coal districts. These situations invariably created stress and tension not only in the labor ranks but also between management and labor. Sometimes these conditions led to violence. Miners in both Buxton and Haydock benefited from the general prosperity in the industry through the stability and continuity of the UMW. Buxton and Haydock remained unionized throughout their existence.

The founding of Buxton also came at a propitious time in regard to welfare capitalism. Although in the broadest sense the construction of educational institutions, churches, YMCAs, and housing projects was designed to improve the quality of life for employees, it was more specifically intended to reduce labor discontent, thus making workers less inclined to support trade unions. The entire planning and development of Buxton can be viewed from the perspective of "enlightened self-interest." Although there is no direct evidence that John or Ben Buxton, or any other Consolidation official, was devoted to carrying out the tenets of welfare capitalism, the company's actions were certainly in accord with that movement. Buxton was laid out with considerable attention to its appearance and functioning. Ben Buxton, the town's major planner, determined that the homes would be spacious and sturdy. His plans, moreover, called for several parks, two tennis courts, and two baseball diamonds. Shortly after Buxton's inception, Consolidation also constructed two YMCAs for its employees. Although contemporaries viewed their construction as altruistically motivated, it can better be viewed as a move in keeping with the tenets of welfare capitalism. But while the YMCAs accomplished the goals of the company, particularly that of reducing worker discontent, the benefits to the employees were far-reaching and must be recognized. Interviews with former Buxton residents indicate that the YMCAs served as the centers of social life in Buxton for all residents, both black and white, and provided leadership and inspiration for young black males.

Consolidation's management policies should be viewed in the context of the prosperous times at the turn of the century. Good profits from the sale of coal made it less necessary for the company to control all other business operations as an additional source of profits. Nevertheless, Consolidation deserves attention for its liberal housing and business practices. While coal operators frequently exercised tight control over the economic life of their towns, Consolidation officials followed many flexible practices. Company officials allowed independent businessmen to operate, with the result that Buxton contained a wide assortment of private stores, hotels, bakeries, and pool halls. In housing, the company followed similarly liberal policies. Individuals could rent land from the company on which to build a house, or they could purchase small acreages in surrounding areas. Unlike many coal operators, Consolidation did not enforce the practice of requiring a man to rent a company house in order to be hired. This meant that both blacks and whites were able to combine mining and farming by purchasing acreages that provided them with additional income. As a result, many Buxton residents lived more comfortably than did miners in other Iowa coal communities. Moreover, the acreages provided families with economic security once the men had retired from mining.

Consolidation also allowed considerable latitude in other community institutions. It is not known if the company donated land for Buxton's churches or contributed money for their support, but there is also no evidence that it took an active part in determining church policies or selecting ministers. David Corbin has pointed out in his study of coal mining in southern West Virginia that coal operators there frequently exercised tight control over their towns, even to the point of hiring preachers and schoolteachers.[5] In Buxton, Consolidation apparently maintained tight control over the management of the company store and bank but allowed other businesses and social and educational institutions to operate independently.

Reuben Gaines, Sr., owner of several businesses and rental homes in Gainesville, a suburb of Buxton. (*Courtesy of Donald Gaines*)

The community that appeared in 1900 was an atypical mining community in many respects. Unlike other Iowa mining camps, which were squeezed in tightly next to the coal tipples, and unlike camps in Pennsylvania or West Virginia, which were perched on mountainsides or located deep in narrow valleys, Buxton offered residents a spacious, attractive environment. Houses were comfortable and well maintained. Residents had ample space, since each house was located on a quarter-acre lot. The company and private businessmen provided residents with an abundance of shopping facilities. Schools and churches were evident throughout the town, and the YMCAs served as the center of a rich and varied social life.

Moreover, throughout most of its existence Buxton prospered. Because Consolidation was a captive mine (in which all coal was purchased by the parent company, the Chicago and North Western Railroad), its employees averaged more working days than employees of many independent Iowa mines, which operated on a seasonal basis. Merchants in Buxton also prospered from the steady wages of the mine employees. The result was that most Buxton residents lived well. In the words of Hucey Hart, "In Buxton, we didn't want for nothing."[6]

Of all Buxton's characteristics, however, the most significant and widely recognized was its harmonious race relations. Black residents viewed Buxton as an almost perfect place to live. All black informants and many white informants commented extensively on that fact. Black residents remembered that the company treated all employees fairly and that blacks and whites mingled both at work and at play. Many black individuals who worked for Consolidation had been slaves before the Civil War and had later worked as southern field hands or sharecroppers. To these people, life in Buxton seemed vastly different from what they had known in the South. Certainly it also differed from what many southern blacks experienced in the industrial cities of the Northeast. For blacks raised in Buxton, however, the most dramatic change came when they left there and settled in urban centers. Informant after informant commented on the rude awakening they experienced after settling in Cedar Rapids, Waterloo, or a larger northeastern city.

From all perspectives, Buxton must be considered a success story for blacks. At a time when racial discrimination was rampant throughout the country, resulting in blacks' inability to find decent jobs, adequate housing, or good school facilities, black families in Buxton

An unidentified young woman who—perhaps posing for a special occasion such as graduation, a church confirmation, or a sweet sixteen party—seems to symbolize the pleasant life in Buxton that former residents remember so well. (*Courtesy of Iowa State Historical Department*)

shared equally in the good life. In Buxton, black families faced no barriers in renting or building homes, in finding employment, or in patronizing public institutions and private businesses. In fact, blacks enjoyed some institutions designed especially for their race. Their experience in Buxton stands in contrast not only to the treatment of blacks in other Iowa coal camps, where they were often excluded or restricted to a particular section of the camp, but also to the treatment of blacks in both the northern and southern states.

Finally, it should be noted that black residents took maximum advantage of the opportunities Buxton provided. The prosperity of the Iowa coal industry allowed for full-time work, and blacks in Buxton responded to that situation, making good wages for many years. The community also provided opportunities for black professionals and black businessmen, and, at least in Buxton's early years, these people dominated the town. Many black families owned their own homes, many of which the *Bystander* cited as among the largest and most comfortable homes in Buxton. Some black families took advantage of good wages to send their sons and daughters to college. Moreover, blacks formed innumerable social, literary, and musical organizations through which they further developed their own talents and added greatly to the quality of life in Buxton.

An assessment of Buxton from the perspective of white residents is more difficult. Probably the greatest advantage for whites resulted from the size of the community. Many white former residents had worked in other Iowa camps in which a population that did not exceed several hundred people was served by a small company store, a pool hall, and the miners' union hall. Buxton, by comparison, with its population of nearly 5,000, offered the shopping facilities and social opportunities characteristic of a substantial Iowa community. For whites as well as blacks, the company's housing and business policies were advantageous. The Swedes especially took advantage of the option to rent land and build their own homes. Numerous whites also established businesses there. The one great contrast between the lives of whites and blacks, however, was in the matter of racial discrimination. While whites frequently faced discrimination outside Buxton because of their mining occupation, they experienced little discrimination because of their race. Perhaps the one exception was with the southern and eastern Europeans who came to Iowa's coalfields. Many of these people experienced discrimination elsewhere because of their origin and because of their Catholic religion. This was more

Poster advertising a Buxton reunion in 1952. (*Courtesy of Iowa State Historical Department*)

subtle, however, than racial discrimination and it did not necessarily result in their being denied housing and jobs. For many white former Buxton residents, the community was a good place to live, but they did not describe it in the same idyllic terms as did blacks.

For many years after Buxton's abandonment, former residents trooped back to the site to hold summer reunions. Participants estimated that some reunions held in the 1930s attracted several thousand people. People came back to greet friends, neighbors, and relatives and to walk over what remained of the old town, reminiscing about their lives there. For blacks in particular, reunions presented an opportunity to relive, if only briefly, the pleasantness of their former lives. At the same time, residents of communities like Waterloo and Des Moines formed Buxton clubs, which provided both the opportunity to keep old acquaintanceships alive and to promote research on the history of Buxton.

Of all the Buxton residents, however, the demise of the community affected the black population most significantly. Buxton had been proclaimed the "king of the mining communities" because of its large black population and the lives they lived there. Even though in its final years Buxton had ceased to be predominantly black, black people remembered vividly the extreme cordiality and prosperity of their lives there. For blacks, their utopia had faded away. Perhaps Marjorie Brown summed up their feelings best when she said: "Buxton was something else. You can imagine how we grieved for it."[7]

NOTES

INTRODUCTION

1. For a full discussion of coal mining in Iowa, see Dorothy Schwieder, *Black Diamonds: Life and Work in Iowa's Coal Mining Communities, 1895–1925* (Ames: Iowa State University Press, 1983).

2. All persons interviewed for the Buxton project stated that race relations were harmonious and that the Consolidation Coal Company had been a fair employer. The interviews (tapes and typed transcripts) are deposited with the Iowa State Historical Department, Des Moines.

3. Joseph Hraba, *American Ethnicity* (Itasca, Ill.: F. E. Peacock Publishers, 1979), p. 268; and Fred A. Shannon, *The Farmer's Last Frontier: Agriculture, 1860–1897,* Vol. 5 of The Economic History of the United States, ed. Henry David (New York: Farrar and Rinehart, 1945), chapter 4.

4. Sig Synnestvedt, *The White Response to Black Emancipation: Second-Class Citizenship in the United States since Reconstruction* (New York: Macmillan Company, 1972), p. 42; and Mary Beth Norton et al., *A People and a Nation: A History of the United States* (Boston: Houghton Mifflin Company, 1982), 2:459.

5. For a complete discussion of the Jim Crow laws in the South, see C. Vann Woodward, *The Strange Career of Jim Crow* (New York: Oxford University Press, 1955).

6. Norton et al., *A People,* 2:460.

7. Leonard Broom and Norval Glenn, *Transformation of the Negro American* (New York: Harper and Row, 1965), pp. 159–60.

8. Elizabeth Pleck, "A Mother's Wages: Income Earning Among Married Italian and Black Women, 1896–1911," in *A Heritage of Her Own: Toward a New Social History of American Women,* ed. Nancy F. Cott and Elizabeth Pleck (New York: Simon and Schuster, 1979), pp. 368–74.

9. Hraba, *American Ethnicity,* pp. 278–79.

10. Robert Athearn, *In Search of Canaan: Black Migration to Kansas 1879–80* (Lawrence: Regents Press of Kansas, 1978), pp. 78–79.

CHAPTER 1

1. *The History of Mahaska County* (Des Moines: Union Historical Company, 1878), pp. 519–20.

2. Ibid., pp. 519–21; Hubert L. Olin, *Coal Mining in Iowa* (Des Moines: State of Iowa, 1965), pp. 44–46; quoted statement is from *The History of Mahaska County,* p. 521.

3. *Iowa Official Register,* 1915–16, pp. 746–47.

4. For a discussion of Iowa's early development and railroad expansion, see Leland Sage, *A History of Iowa* (Ames: Iowa State University Press, 1974), chapter 6.

5. Charles Keyes, "Historical Sketch of Mining," *Iowa Geological Survey* 22 (1912): 119; and James H. Lees, "History of Coal Mining in Iowa," *Iowa Geological Survey* 19 (1908): 525–33.

6. *Biennial Report of the State Mine Inspectors, 1884–1885* (Des Moines: State of Iowa, 1885), p. 4.

7. See Schwieder, *Black Diamonds,* Chapter 3, for a full discussion of Iowa's coal camps.

8. Ibid.

9. Olin, *Coal Mining,* p. 46.

10. Ibid., p. 49; and Iowa, Census of Iowa, 1885 and 1895, Manuscript Population Schedules for Harrison and East Des Moines Townships, Mahaska County.

11. Olin, *Coal Mining,* pp. 49–53.

12. Leola Nelson Bergmann, *The Negro in Iowa* (Iowa City: State Historical Society of Iowa, 1969), pp. 41–42; and *A History of Lucas County* (Des Moines: State Historical Co., 1881), pp. 611–15. Although black strikebreakers were imported into Iowa for the first time in 1880, Swedes had been imported into Beacon as strikebreakers in 1873. In that year, George Williams, who operated a coal mine in the vicinity of Beacon, decided to open a company store, and he expected all of his employees to trade there. The following spring Williams discharged some miners, including those who had failed to trade at the company store. The remaining miners went out on strike to protest this action, claiming that they should have the right to trade anywhere they wished. The company responded by importing a number of Swedes who had been working on a federal project at Keokuk. This action immediately broke the strike. During the remainder of the 1880s and the 1890s, Iowa coal operators continued to import blacks as strikebreakers. In June 1891, Mystic coal operators followed the same practice, only to have the striking white miners react by greeting the blacks with loaded guns. Shots were exchanged but no one was injured. At the same time, operators imported fifty blacks into the coal camp of Carbonado, where violence also erupted. Intimidated, some of the blacks returned to the South while others endured the harassment and became regular company employees. *History of Mahaska County,* pp. 510–20; Bergmann, *The Negro in Iowa,* p. 42.

13. *Territorial Census of Iowa,* 1840, p. 2.

14. Bergmann, *The Negro in Iowa,* p. 9.

15. Ibid., pp. 12–14; and Sage, *History of Iowa,* p. 83.

16. Joseph Frazier Wall, *Iowa: A Bicentennial History* (New York: W. W. Norton and Company; Nashville: American Association for State and Local History, 1978), p. 115; and Bergmann, *The Negro in Iowa,* p. 54.

17. Bergmann, *The Negro in Iowa,* pp. 37–40.

18. From a table in the *Iowa Official Register,* 1915–1916, pp. 746–47; and Bergmann, *The Negro in Iowa,* p. 41.

19. R. L. Polk and Company, *Iowa State Gazetteer and Business Directory,* 1884–1885, pp. 748–49.

20. Anna Olson to Hubert Olin, March 1963, Olin Papers, Iowa State Historical Department, Des Moines; Minnie B. London, "As I Remember," 1940, Iowa State Historical Department, Des Moines. The Olin Papers were collected between 1960 and 1963 and include material and correspondence related to Iowa coal mining.

21. Polk, *Gazetteer,* 1882–1883, p. 587; David McLaughlin, "Mahaska County Once Led State in Coal Mining Activities," *Oskaloosa Daily Herald,* January 30, 1940. The McLaughlin article is a reprint from a report on coal mining prepared by the Iowa Writer's Program, a unit of the federal government's Work Projects Administration. David McLaughlin was the supervisor.

22. Interviews with Herman Brooks, former resident of Muchakinock, Chicago, Illinois, July 30, 1981; Bessie Lewis, former resident of Muchakinock, Des Moines,

Iowa, January 17, 1981; and Jeanette Adams, former resident of Muchakinock, Des Moines, Iowa, June 21, 1980; and *Iowa State Bystander* (Des Moines), June 18, 1897 (hereafter cited as *Bystander*).

23. Interview with Bessie Lewis.

24. Iowa, Census, 1880 and 1895, Harrison and East Des Moines Townships, Mahaska County.

25. These reports appeared frequently in the *Bystander* between 1896 and 1900. Interviews with both Muchakinock and Buxton residents also confirmed that Consolidation continued to recruit black workers from Virginia throughout the 1880s and 1890s.

26. Interview with Bessie Lewis.

27. Interview with Mattie Murray, former resident of Muchakinock, Des Moines, Iowa, June 4, 1981.

28. Ibid.

29. Interview with Jeanette Adams.

30. Interview with Herman Brooks.

31. Interview with Alex Erickson, former resident of Muchakinock, Pershing, Iowa, August 15, 1980.

32. Ibid.

33. Interview with Andrew Smith, former Buxton resident, Lovilia, Iowa, June 17, 1981. Andrew had changed his name from Mital to Smith.

34. Letter from Clayborne Carter to Hubert Olin, March 1963, Olin Papers.

35. Ibid.; interview with Carl Kietzman, former resident of Buxton and son-in-law of Hobe Armstrong, rural Albia, Iowa, May 28, 1981; and *Eddyville Tribune*. September 15, 1898.

36. Interview with Carl Kietzman.

37. London, "As I Remember," p. 2.

38. Interview with Alex Erickson.

39. Clayborne Carter to Hubert Olin, March 1963, Olin Papers; interview with Earl Smith, former Buxton resident, Albia, Iowa, January 4 and July 13, 1981.

40. Interview with Carl Kietzman.

41. Newspaper article from the *Wytheville* (Va.) *Enterprise*, December 1898; Olin Papers; *Oskaloosa Times*, January 15, 1898; *Bystander*, June 18, 1898.

42. *Bystander*, December 23, 1898, and May 19 and October 13, 1899; and interview with Herman Brooks.

43. McLaughlin, "Mahaska County Once Led State in Coal Mining Activities"; London, "As I Remember," p. 2; Clayborne Carter to Hubert Olin, March 1963, Olin Papers; *Negro Solicitor*, January 20, 1898.

44. *Oskaloosa Weekly Herald*, October 4, 1883.

45. *Oskaloosa Daily News*, March 27, 1897.

46. *Bystander*, July 30, 1896.

47. *Eddyville Tribune*, September 8, 1898; clipping from *Muchakinock State*, December 17, 1897, in scrapbook in the possession of Robert Smith, Bussey, Iowa. About a hundred clippings from the *Solicitor* are contained in the scrapbook.

48. *Bystander*, June 30, July 29, and September 1, 1899.

49. Ibid., May 21, 1897.

50. Ibid., February 3 and 10, 1899.

51. Ibid., July 7 and 14, 1899.

52. *Solicitor*, January 13, 1898.

53. Ibid. In a later issue, the *Solicitor* editor charged: "The Negro Republican contingent of Polk County made a savage, ungraceful fight against Mr. Woodson . . . [which] exposed the unfair methods of the Polk County leaders to such an extent that the Negroes of other parts of the state will in the future array themselves against them" (January 20, 1898).

54. *Bystander,* August 4, 1899.

55. Ibid., July 21, 1899.

56. *Oskaloosa Weekly Herald,* June 7, 1883.

57. By 1895, Muchakinock's population had increased considerably. No doubt some of this resulted from an increase in the population of the town itself, but some of the increase also came from the creation of other coal camps nearby. In coding census data for unincorporated communities like Muchakinock, it is necessary to include all residents of the township in which the camp was located, because the census taker did not separate the camp's population from the remainder of the township population. In the case of Muchakinock, the camp straddled the line between two townships, Harrison and East Des Moines. Because of this, it was necessary to include all residents in the two townships in coding the data. It is known that at least five coal camps came into existence in Harrison and East Des Moines Townships between 1885 and 1895.

58. Anton Lundeen, *Iowa-Konferenses Af Augustana-Synoden Historia* (Augustana Conference, n.d.), translation by Karen Cleven, Madrid, Iowa, in possession of Mr. and Mrs. Norman Bergstrom, rural Ogden, Iowa.

59. Ibid.

60. Ibid.

61. Iowa, Census, 1885 and 1895, Harrison and East Des Moines Townships, Mahaska County.

62. Interview with Herman Brooks.

63. *Solicitor,* July 9, 1898.

64. *Bystander,* June 18, 1897.

CHAPTER 2

1. Iowa, Census of Iowa, 1905, Manuscript Population Schedules for Bluff Creek Township, Monroe County.

2. David McLaughlin, "Mahaska County Once Led State in Coal Mining Activities," *Oskaloosa Daily Herald,* January 30, 1940.

3. Consolidation Coal Company, Minutes of Business Meetings, vol. 2, June 30, 1896; interview with Ada Morgan, daughter of the general superintendent, E. M. Baysoar, and former resident of Buxton, Peru, Illinois, July 2, 1981.

4. Letter from Clayborne Carter to Hubert Olin, February 26, 1963, Olin Papers. Mr. Carter was a brother of Dr. E. A. Carter and a former resident of Buxton.

5. Charles Nichols, *Housing Conditions: Iowa Coal Mining Camps* (Des Moines: State Printing Office, 1919), p. 20.

6. McLaughlin, "Mahaska County"; Reuben Gaines, Jr., "Memoirs of Buxton," 1980, p. 52, in possession of Donald Gaines, rural Albia, Iowa.

7. Interviews with Herman Brooks, former resident of Muchakinock and Buxton, Chicago, Illinois, July 30, 1981; and Bessie Lewis, former resident of Muchakinock and Buxton, Des Moines, Iowa, January 17, 1981; *Iowa State Bystander* (Des Moines), March 18, 1903.

8. Interviews with Jacob Brown, former resident of Buxton, Waterloo, Iowa, November 18, 1980; Charles Lenger, former resident of Buxton, Oskaloosa, Iowa, July 22, 1981; Mike Onder, former resident of Buxton, Albia, Iowa, June 23, 1981; Lester Beaman, former resident of Buxton, Waterloo, Iowa, November 17, 1980; and Carl Kietzman, former resident of Buxton, rural Albia, Iowa, May 28, 1981; and *Bystander,* December 29, 1916. John E. Baxter, a former Buxton resident, estimated in a speech that in its first decade Buxton contained approximately 2,000 homes. It would seem from other accounts that Baxter's estimate is excessive. Baxter's speech is in the John E. Baxter correspondence in the Olin Papers. Copies of plat map in possession of Dorothy

Collier and other Buxton Club members, Des Moines, Iowa. No date given.

9. Interviews with Dorothy Collier, former Buxton resident, Des Moines, Iowa, October 17, 1980; Vaeletta Fields, former Buxton resident, Waterloo, Iowa, July 7 and 16, 1981; and Charles Lenger.

10. Interviews with Jacob Brown; Helen Duke, former Buxton resident, Des Moines, Iowa, June 30, 1981; and Dorothy Collier.

11. Interviews with Marjorie Brown, former Buxton resident, Waterloo, Iowa, August 19 and 20, 1980; and Dorothy Collier.

12. Interviews with Susie Robinson, former Buxton resident, Des Moines, Iowa, June 4, 1981; Herman Brooks; Mike Onder; and Lara Wardelin, former Buxton resident, Des Moines, Iowa, July 8, 1981.

13. Interviews with Hucey Hart, former Buxton resident, Des Moines, Iowa, November 1, 1980; Lester Beaman; and Hazel Stapleton, former resident of Buxton, Des Moines, Iowa, June 24, 1980.

14. Interviews with Bessie Lewis and Susie Robinson.

15. Interview with Gertrude Stokes, former Buxton resident, Cedar Rapids, Iowa, June 15, 1981.

16. Interview with Agnes and Alex Erickson, former residents of Buxton, Pershing, Iowa, August 15, 1980; and *Bystander,* September 18, 1903.

17. Interviews with Harvey Lewis, former Buxton resident, Bondurant, Iowa, December 29, 1980; and Archie Harris, former resident of Buxton, rural Lovilia, Iowa, November 15, 1980.

18. *Bystander,* November 15, 1901; map of Buxton drawn by Clayborne Carter, Olin Papers; and interviews with Ada Morgan; and Marvin Franzen, former resident of Buxton, Lovilia, Iowa, September 9, 1981.

19. *Bystander,* November 15, 1902; and interview with Irene Goodwin, former resident of Buxton, Lovilia, Iowa, June 28, 1983.

20. *Bystander,* June 21, 1901; Gaines, "Memoirs of Buxton," p. 51; and interview with Archie Harris.

21. Interview with Leroy Wright, former Buxton resident, Waterloo, Iowa, November 18, 1980.

22. Interview with Herman Brooks; speech by John E. Baxter, Ottumwa, Iowa, for presentation at Buxton Association, Des Moines, September 1963, Olin Papers. Baxter was a former resident of Buxton, Interviews with Mike Onder and Agnes and Alex Erickson.

23. Interviews with Archie Harris, Agnes and Alex Erickson, and Lara Wardelin; and *Bystander,* October 17, 1907, and October 16, 1903.

24. Interviews with Leroy Wright; Mike Onder; Clifford Lewis, former resident of Buxton, Eddyville, Iowa, May 28, 1981; and George Klobnak, former Buxton resident, Albia, Iowa, June 18, 1981.

25. Interviews with Clifford Lewis; George Klobnak; Mike Onder; and Earl Smith, former Buxton resident, Albia, Iowa, July 13, 1981.

26. Interviews with Jacob Brown; Mike Onder; Bessie Lewis; and Lola Reeves, former Buxton resident, Des Moines, Iowa, June 21, 1980.

27. Stanley Miller, "As Others See Buxton," *Mt. Pleasant Free Press,* December 1910.

28. John Baxter speech, Olin Papers.

29. *Bystander,* November 17, 1905.

30. Ibid., December 6, 1907; and interview with Bessie Lewis.

31. *Bystander,* December 6, 1907.

32. Interviews with Carl Kietzman and Gertrude Stokes.

33. *Bystander,* December 6, 1907.

34. Ibid.

35. Ibid.; and interviews with Alex Erickson and Mike Onder.

36. Interview with Carl Kietzman.

37. *Bystander,* December 6, 1907; and interview with Sister Maurine Sofranko, former resident of rural Buxton, Ottumwa, Iowa, September 28, 1980; and interviews with Dorothy Collier and Hazel Chapman.

38. Interview with Gertrude Stokes.

39. Interview with Robert Wheels, former resident of Buxton, Des Moines, Iowa, June 8, 1981.

40. *Bystander,* November 17, 1905.

41. *Bystander,* March 3, April 21, and June 23, 1911.

42. *Bystander,* October 20, 1911.

43. Interview with Odessa Booker, former resident of Buxton, Des Moines, Iowa, June 8, 1981.

44. Interviews with Susie Robinson, Agnes Erickson, and Bessie Lewis; letter from Clayborne Carter to Hubert Olin, February 26, 1963, Olin Papers.

45. Interview with Mike Onder.

46. R. L. Polk and Company, *Iowa State Gazetteer and Business Directory* (Des Moines: R. L. Polk and Co., 1906), p. 25; *Bystander*, December 6, 1907; and interview with Mike Onder.

47. Interviews with Walter and Mildred Gardner, former residents of Buxton, Albia, Iowa, June 4, 1981; and Archie Allison, former resident of Buxton, Albia, Iowa, June 18, 1981.

48. Interviews with Harold Reasby, former resident of Buxton, Waterloo, Iowa, July 21, 1981; and Mike Onder.

49. *Bystander,* November 27, 1903.

50. Ibid.; Minnie B. London, "As I Remember," 1940, p. 1, Iowa State Historical Department, Des Moines; *Albia Republican*, March 10, 1910; and Clayborne Carter to Hubert Olin, February 26, 1963, Olin Papers.

51. Interview with Carl Kietzman; *Bystander,* November 25, 1910.

52. *Bystander,* April 25, 1902, and July 19, 1903.

53. *Bystander,* October 20, 1905, and December 6, 1907.

54. Ibid.

55. Ibid., August 29, 1902.

56. U.S. Department of Commerce, Bureau of the Census, *Thirteenth Census of the United States, 1910,* Manuscript Population Schedules for Bluff Creek Township, Monroe County, Iowa.

57. *Albia Republican,* February 10, 1910. The estimate of a thousand employees in 1910 is from the *Republican*. It is impossible to know the exact number of employees because of the lack of company records.

58. Interviews with Archie Harris, Alex Erickson, and Bessie Lewis.

CHAPTER 3

1. *Iowa State Bystander* (Des Moines), September 18, 1903.

2. Interview with Archie Harris, former resident of Buxton, rural Lovilia, Iowa, November 15, 1980.

3. In 1880 the Iowa General Assembly passed legislation that prohibited boys under twelve from working underground; *Laws of Iowa,* 1880, p. 199.

4. Interviews with Hucey Hart, former resident of Buxton, Des Moines, Iowa, November 1, 1980; and Alex Erickson, former resident of Buxton, Pershing, Iowa, August 15, 1980.

5. Interviews with Mattie Murray, former resident of Buxton, Des Moines, Iowa,

June 4, 1981; and Hubert L. Olin, *Coal Mining in Iowa* (Des Moines: State of Iowa, 1965), p. 49.

6. United Mine Workers of America, District 13, Joint Wage Agreement, 1910–1912, p. 17, Iowa State Historical Department, Des Moines; interview with Robert Wheels, former resident of Buxton and Haydock, Des Moines, Iowa, June 8, 1981. New employees who chose to work as company men (men who performed specialized work, such as timbering, tracklaying, or mule driving) were also hired as helpers. In 1910 the Joint Wage Agreement listed the wage of helpers for tracklayers and timbermen as $2.49 per day.

7. Iowa, Census of Iowa, 1905 and 1915, Manuscript Population Schedules for Bluff Creek Township, Monroe County.

8. Iowa, *Biennial Report of the State Mine Inspectors, 1910* (Des Moines: State of Iowa, 1910), pp. 46–47.

9. Interviews with Reuben Gaines, Jr., former resident of Buxton, rural Albia, Iowa, February 19, 1981; and Sister Maurine Sofranko, former resident of rural Buxton, Ottumwa, Iowa, September 28, 1980.

10. Donald Baker, "Operating Conditions Encountered in Iowa," *Coal Age* 14 (March 1921): 437.

11. Interview with Archie Harris.

12. Interview with Milo Papich, former mine foreman, Slater, Iowa, May 8, 1979.

13. *Albia Republican*, March 3, 1910; Iowa, Census, 1915, Bluff Creek Township. Because the census was taken at midyear, in June 1915, census takers were instructed to ask residents what they had earned in the previous full year, 1914.

14. *Bystander*, December 6, 1907; and interviews with Robert Wheels; Hucey Hart; Jacob Brown, former resident of Buxton, Waterloo, Iowa, November 18, 1980; and Herman Brooks, former resident of Muchakinock and Buxton, Chicago, Illinois, July 30, 1981.

15. Iowa, Census, 1915, Bluff Creek Township; *Bystander*, January 9 and June 26, 1914.

16. Iowa, Census, 1915, Manuscript Population Schedules for Beacon, Cincinnati, and Seymour.

17. Iowa, Census, 1915, Bluff Creek Township.

18. Ibid.; and interview with Agnes and Alex Erickson, former residents of Buxton, Pershing, Iowa, August 15, 1980.

19. Iowa, Census, 1915, Bluff Creek Township.

20. Ibid.

21. UMW, District 13, Joint Wage Agreement, 1910–1912, p. 17. District 13 was divided into four subdistricts, with slightly different wage scales in each subdistrict because of physical differences among them. District 13 included all of Iowa plus Putnam County, Missouri.

22. Interviews with Alex Erickson and Robert Wheels.

23. Interview with Alex Erickson.

24. Ibid.; interviews with Hucey Hart and Jacob Brown.

25. Interview with Dean Aubrey, former state mine inspector, Ames, Iowa, August 2, 1972.

26. For a full discussion of the United Mine Workers in Iowa, see Dorothy Schwieder, *Black Diamonds: Life and Work in Iowa's Coal Mining Communities, 1895–1925* (Ames: Iowa State University Press, 1983), chapter 6; United Mine Workers of America, District 13, Executive Board Proceedings, 4:8, Iowa State Historical Department, Des Moines.

27. Schwieder, *Black Diamonds*, pp. 130–40.

28. Ibid., pp. 140–41; and interview with Odessa Booker, former resident of Buxton, Des Moines, Iowa, June 8, 1981.

29. *United Mine Workers Journal,* August 5, 1909; December 30, 1915; and January 20, 1916; and interview with Nellie Lash King, former resident of Buxton, Lovilia, Iowa, December 23, 1980.

30. *Bystander,* December 6, 1907.

31. Interview with Herman Brooks.

32. Ibid.

33. Iowa, Census, 1915, Bluff Creek Township; and Schwieder, *Black Diamonds,* pp. 146–50.

34. *Bystander,* January 26, 1912; and UMW, District 13, Executive Board Proceedings, 2:1, 2.

35. UMW, District 13, Executive Board Proceedings, 1:78.

36. Ibid., 1:91.

37. William Ewalt, "The Ogden Coal Strike, 1910–1912," (Student paper, Iowa State University, 1975), pp. 1–20, in possession of Dorothy Schwieder.

38. Ibid., pp. 20–25.

39. Interviews with Alex Erickson and Hucey Hart.

40. Interview with Alex Erickson.

41. Quoted in A. T. Shurick, *The Coal Industry* (Boston: Little, Brown, and Company, 1924), p. 311; and *Coal Age* 8 (November 1915): 895.

42. Interview with Lola Nizzi, Granger, Iowa, January 10, 1978.

43. Carter Goodrich, *The Miner's Freedom: A Study of the Working Life in a Changing Society* (Boston: Marshall Jones Company, 1925), pp. 99–100.

44. Interviews with Sister Maurine Sofranko and Mary Allison, former resident of rural Buxton, Albia, Iowa, February 19, 1981.

45. Interview with Lola Reeves, former resident of Buxton, Des Moines, Iowa, June 21, 1980.

CHAPTER 4

1. Stuart D. Brandes, *American Welfare Capitalism, 1880–1940* (Chicago: University of Chicago Press, 1976), pp. 5–6, 14–15.

2. Ibid., pp. 14–16.

3. Ibid., pp. 16–19; and Ray Ginger, *Altgeld's America: The Lincoln Ideal Versus Changing Realities* (New York: Funk and Wagnalls Company, 1958), pp. 145–47.

4. Brandes, *American Welfare Capitalism,* pp. 33, 34, 81.

5. It has not been possible to locate any business records other than a few sets of minutes from the Consolidation Coal Company board of directors meetings in Chicago in the late 1800s and early 1900s. These generally give only the order of business and do not include minutes of discussions or even the board's decisions. Moreover, it has not been possible to locate any private correspondence of the Buxton family or to locate any of their descendents.

6. Brandes, *American Welfare Capitalism,* p. 46. Virtually all of the seventy-five people interviewed commented on the pleasant and extensive facilities created in Buxton by Consolidation.

7. *United Mine Workers Journal,* September 20, 1900.

8. Interviews with Hucey Hart, former resident of Buxton, Des Moines, Iowa, November 1, 1980; and Elmer Buford, former resident of Buxton, Des Moines, Iowa, June 30, 1980.

9. For a discussion of Iowa coal camps, see Dorothy Schwieder, *Black Diamonds: Life and Work in Iowa's Coal Mining Communities, 1895–1925* (Ames: Iowa State University Press, 1983), Chapter 3.

10. Almost every informant commented on the fact that houses in Buxton were

superior to houses in other coal camps.

11. See Chapter 2 for a complete discussion of buildings and facilities.

12. Charles Nichols, *Housing Conditions: Iowa Coal Mining Camps* (Des Moines: State Printing Office, 1919), p. 31.

13. John E. Baxter Speech, Olin Papers.

14. Interviews with Alex Erickson, former resident of Buxton, Pershing, Iowa, August 15, 1980; and Jessie Frazier, former resident of Buxton, Des Moines, Iowa, March 15, 1979; and *Iowa State Bystander* (Des Moines), December 21, 1909, and July 7 and August 9, 1911.

15. *Bystander,* September 19, 1903, and January 13, 1911.

16. Interview with Robert Wheels, former resident of Buxton, Des Moines, Iowa, June 8, 1981.

17. Interview with Susie Robinson, former resident of Buxton, Des Moines, Iowa, June 4, 1981.

18. Interview with Odessa Booker, former resident of Buxton, Des Moines, Iowa, June 8, 1981.

19. Ibid.

20. Interview with Adolph Larson, former resident of Buxton, Knoxville, Iowa, July 23, 1981.

21. *Bystander,* December 6, 1907.

22. Ibid.

23. David Corbin, *Life, Work, and Rebellion in the Coal Fields: The Southern West Virginia Miners, 1880–1922* (Urbana: University of Illinois Press, 1981), pp. 68–70.

24. Brandes, *American Welfare Capitalism,* pp. 66, 71.

25. Interview with Irene Goodwin, former resident of Buxton, Lovilia, Iowa, June 28, 1983.

26. Newspaper article from the *Wytheville* (Va.) *Enterprise,* December, 1899, Olin Papers; *Oskaloosa Times,* January 15, 1898; *Bystander,* May 21, 1897; Corbin, *Life, Work, and Rebellion,* pp. 68–70.

27. *Bystander,* January 8, 1909; December 16 and 30, 1910; and January 13, 1911; Clayborne Carter to Hubert Olin, March 1963, Olin Papers.

28. Interview with Earl Smith, former Buxton area resident, Albia, Iowa, January 4, 1981.

29. Interview with Hucey Hart.

30. Interviews with Dorothy Collier, former Buxton resident, Des Moines, Iowa, October 17, 1980; and Gertrude Stokes, former Buxton resident, Cedar Rapids, Iowa, June 15, 1981.

31. Interview with Bessie Lewis, former resident of Buxton, Des Moines, Iowa, January 17, 1981.

32. Ibid.

33. Corbin, *Life, Work, and Rebellion,* pp. 125–26.

34. *Bystander,* September 18, 1903.

35. Ibid., October 24, 1913.

36. Ibid., July 15, 1912; and interview with Marjorie Brown, former Buxton resident, Waterloo, Iowa, August 19 and 20, 1980.

37. Interviews with Jacob Brown, former resident of Buxton, Waterloo, Iowa, November 19, 1980; and Mike Onder, former resident of Buxton, Des Moines, Iowa, June 23, 1981.

38. Interviews with Leroy Wright, former resident of Buxton, Waterloo, Iowa, November 18, 1980; and Harold Reasby, former resident of Buxton, Waterloo, Iowa, July 21, 1981; and *Bystander,* October 24, 1913.

39. *Bystander*, February 24, 1905.

40. *Bystander,* November 15, 1918.

41. Interview with Robert Wheels.

42. *Buxton Eagle,* April 27, 1905. This editorial is a clipping in the collection of Donald Gaines, rural Albia, Iowa.

43. Ibid. It is not clear why the *Eagle* editor accused the company of ruining the educational principle of Buxton by maintaining a separate white school. The only all-white school in Buxton was operated by the Swedes for only a few years.

CHAPTER 5

1. Interviews with Agnes Erickson, former resident of Buxton, Pershing, Iowa, August 15, 1980; and Irene Goodwin, former resident of Buxton and Haydock, Lovilia, Iowa, June 28, 1983. Agnes Erickson and many other informants stated that married women rarely worked outside the home in Buxton; rather, they took care of the house.

2. Interviews with Jacob Brown, former resident of Buxton, Waterloo, Iowa, November 18, 1980; and Charles Lenger, former resident of Buxton, Oskaloosa, Iowa, July 22, 1981.

3. Interview with Bessie Lewis, former resident of Buxton, Des Moines, Iowa, January 17, 1981.

4. Interview with Gertrude Stokes, former resident of Buxton, Cedar Rapids, Iowa, June 15, 1981.

5. Interviews with Lara Wardelin, former resident of Buxton, Des Moines, Iowa, July 8, 1981; and Jacob Brown.

6. Interviews with Dorothy Collier, former resident of Buxton, Des Moines, Iowa, October 17, 1980; and Agnes Erickson.

7. Interviews with Mattie Murray, former resident of Buxton and Muchakinock, Des Moines, Iowa, June 4, 1981; Robert Wheels, former resident of Buxton and Haydock, Des Moines, Iowa, June 8, 1981; Odessa Booker, former resident of Buxton, Des Moines, Iowa, June 8, 1981; Hucey Hart, former resident of Buxton and Haydock, Des Moines, Iowa, November 1, 1980; Oliver Burkett, former resident of Buxton, Waterloo, Iowa, November 17, 1980; and Susie Robinson, former resident of Buxton, Des Moines, Iowa, June 4, 1981. While most censuses can be used to obtain family data, such as the number of children per family and the number of boarders per family, the state censuses of both 1905 and 1915 present numerous difficulties for researchers. These two censuses were not recorded with households or families as a unit, as were the state censuses of 1855 through 1895. Rather, they were placed on cards (with one card per person) and then filed in alphabetical order. There is no data on the individual cards to allow the researcher to connect them so that households can be recreated. It is possible to arrange the data in households, but these arrangements are somewhat arbitrary. For this reason, we have not tabulated the number of children per household. On the other hand, individual data can be taken from the census of 1905 and 1915 with no reservations.

8. Interview with Gertrude Stokes.

9. Interview with Mattie Murray.

10. Ibid.

11. Interviews with Odessa Booker, Bessie Lewis, and Oliver Burkett.

12. Monroe County, Iowa, Bluff Creek Township Birth Records, 1904–1905. These records were collected by the Iowa State Board of Health and are now located in the Monroe County Sheriff's Office, Albia, Iowa.

13. Interviews with Jacob Brown and Bessie Lewis.

14. Interviews with Agnes Erickson and Irene Goodwin; Irene Goodwin was the only interviewee to mention Indians.

15. Interviews with Susie Robinson, Mattie Murray, Robert Wheels, and Gertrude Stokes.

16. Interviews with Bessie Lewis and Irene Goodwin.
17. Interview with Susie Robinson.
18. Interviews with Susie Robinson and Odessa Booker.
19. Interview with Odessa Booker.
20. Interview with Wilma Stewart, former resident of Buxton, Albia, Iowa, July 21, 1981.
21. Interviews with Hazel Stapleton, former resident of Buxton, Des Moines, Iowa, June 24, 1980; Charles Lenger; and Hucey Hart.
22. Interviews with Lester Beaman, former resident of Buxton, Waterloo, Iowa, November 17, 1980; Earl Smith, former resident of Buxton, Lovilia, Iowa, July 13, 1981; and Oliver Burkett.
23. Interviews with Helen Duke, former resident of Buxton, Des Moines, Iowa, June 30, 1981; and Lester Beaman.
24. Interviews with Dorothy Collier and Odessa Booker.
25. Interview with Gertrude Stokes.
26. Interviews with Nellie Lash King, former resident of Buxton, Lovilia, Iowa, December 23, 1980; Charles Lenger; and Alex and Agnes Erickson.
27. See Dorothy Schwieder, *Black Diamonds: Life and Work in Iowa's Coal Mining Communities, 1895–1925* (Ames: Iowa State University Press, 1983), chapter 4.
28. Interviews with Agnes Erickson and Bessie Lewis.
29. John Bodnar, Roger Simon, and Michael Weber, *Lives of Their Own: Blacks, Italians, and Poles in Pittsburgh, 1900–1960,* The Working Class in American History Series (Urbana: University of Illinois Press, 1982), p. 103.
30. Iowa, Census of Iowa, 1905 and 1915, Manuscript Population Schedules for Bluff Creek Township, Monroe County.
31. Ibid.
32. Ibid.
33. Ibid.
34. Ibid.
35. Interview with Dorothy Collier; and Schwieder, *Black Diamonds,* pp. 114–15.
36. Iowa, Census, 1915, Bluff Creek Township, Monroe County.
37. Ibid.
38. Interview with Agnes Erickson; Iowa, Census, 1915, Bluff Creek Township, Monroe County. After the Ericksons moved to Pershing in 1923, Agnes worked in the company office of the Pershing Coal Company.
39. Interviews with Nellie Lash King, Bessie Lewis, and Gertrude Stokes.
40. Interview with Odessa Booker.
41. Herbert G. Gutman, *The Black Family in Slavery and Freedom, 1750–1925* (New York: Vintage Books, 1977), pp. 541, 608; and Elizabeth Pleck, "A Mother's Wages: Income Earning Among Married Italian and Black Women, 1896–1911," in *A Heritage of Her Own: Toward a New Social History of American Women,* ed. Nancy F. Cott and Elizabeth Pleck (New York: Simon and Schuster, 1979), pp. 368, 374.
42. For a full discussion of male wages in Buxton, see Chapter 3. There are several sources that provide a comparison with wages in Buxton. Margaret Byington found that of ninety households in Homestead, Pennsylvania, twenty-four men earned less than twelve dollars per week, while forty-three earned between twelve and twenty dollars per week. The remainder earned twenty dollars or more. According to a study made by Frank Sheridan, Italian workmen in New York City made $1.46 per day, while Slavic and Hungarian immigrants made slightly higher wages. In 1905 and 1906, three railroads in New York, Pennsylvania, and New Jersey estimated that Italian laborers averaged $37.07 per month. By comparison, men working for Consolidation in Buxton in 1910 as tracklayers, timbermen, and mule drivers received $2.70 per day. Margaret Byington, *Homestead: The Households of a Mill Town,* The Pittsburgh Survey, Vol. 4 (New York: Russell Sage Foundation, 1910), p. 38; Frank Sheridan, "Italian, Slavic and

Hungarian Unskilled Immigrant Laborers in the United States," *Labor Bureau Bulletin* 15 (1907), 406, 473–78; United Mine Workers of America, District 13, 1910–1912 Joint Wage Agreement, p. 17, Iowa State Historical Department, Des Moines.

43. Interviews with Robert Wheels and Bessie Lewis.

44. Interview with Herman Brooks, former resident of Muchakinock and Buxton, Chicago, Illinois, July 30, 1981.

45. Bodnar, Simon, and Weber, *Lives of Their Own,* p. 92.

46. Ibid., pp. 60, 92.

47. Interview with Mattie Murray.

48. Interviews with Archie Harris, former resident of Buxton, rural Lovilia, Iowa, November 15, 1980; Sister Maurine Sofranko, former resident of Buxton, Ottumwa, September 28, 1980.

49. Interview with Alex and Agnes Erickson.

50. Interview with Irene Goodwin.

51. Interviews with Susie Robinson and Dorothy Collier.

52. Interview with Bessie Lewis.

53. Interview with Odessa Booker.

54. Interviews with Oliver Burkett and Lester Beaman.

55. Interviews with Mattie Murray and Odessa Booker.

56. Interviews with Mattie Murray, Dorothy Collier, and Lester Beaman.

57. Interviews with Oliver Burkett and Dorothy Collier.

58. Interview with Mattie Murray.

59. Interviews with Susie Robinson, Mattie Murray, and Gertrude Stokes.

60. Interviews with Dorothy Collier, Marjorie Brown, and Hazel Stapleton.

61. Interviews with Odessa Booker, Lara Wardelin, and Mattie Murray.

62. Interviews with Robert Wheels, Hucey Hart, Oliver Burkett, Lester Beaman, and Jacob Brown.

63. Interviews with Odessa Booker; and Harold Reasby, former resident of Buxton, Waterloo, Iowa, July 21, 1981.

64. Interview with Dorothy Collier, November 14, 1978.

65. Interview with Vaeletta Fields, former resident of Buxton, Waterloo, Iowa, July 7, 1981.

66. Interview with Marjorie Brown, former resident of Buxton, Waterloo, Iowa, October 15, 1978.

67. *Iowa State Bystander* (Des Moines), September 26 and February 27, 1902; and July 10, 1903.

68. Ibid., August 21, 1903.

69. Interview with Odessa Booker.

70. Interview with Gertrude Stokes.

CHAPTER 6

1. *Iowa State Bystander* (Des Moines), August 29, 1902; October 20, 1905; and November 4, 1910.

2. Iowa, Census of Iowa, 1905 and 1915, Manuscript Population Schedules for Bluff Creek Township, Monroe County.

3. The Slovaks were listed in the censuses as having been born in Austria or Hungary. They are grouped together in this study and listed as Slovaks because they were recognized as people from the area of Slovakia, which was controlled by the Austro-Hungarian Empire before World War I. Since then, it has been part of Czechoslovakia. Slovaks are considered a distinct ethnic group among the Slavs having their own language. The Slovaks in Buxton defined themselves as Slovaks, as is clearly evident in the oral histories.

4. Iowa, Census, 1905 and 1915, Bluff Creek Township.
5. *Bystander,* February 6, 1914.
6. Interview with Hucey Hart, former resident of Buxton, Des Moines, Iowa, November 1, 1980.
7. Interviews with Elmer Buford, former resident of Buxton, Des Moines, Iowa, June 30, 1980; and Odessa Booker, former resident of Buxton, Des Moines, Iowa, June 8, 1981.
8. Interview with Marjorie Brown, former resident of Buxton, Waterloo, Iowa, August 19, 1980; *Bystander,* September 26 and April 10, 1902; interview with Carl Kietzman, former resident of Buxton, rural Albia, Iowa, May 28, 1981.
9. Interview with Hucey Hart; *Bystander,* March 19, September 26, and October 24, 1902; and January 11, 1907.
10. *Bystander,* February 28, 1902, and January 7, 1910.
11. Ibid., October 10 and 17, 1902.
12. Interview with Herman Brooks, former resident of Buxton, Chicago, Illinois, July 30, 1981.
13. Interview with Dorothy Collier, former resident of Buxton, Des Moines, Iowa, October 17, 1980.
14. *Bystander,* August 11, 1911, and December 17, 1902.
15. Interviews with Herman Brooks and Elmer Buford.
16. Interviews with Bessie Lewis, former resident of Buxton, Des Moines, Iowa, January 17, 1981; and Elmer Buford.
17. *Bystander,* June 13, 1902.
18. Interviews with Mike Onder, former resident of Buxton, Des Moines, Iowa, June 23, 1981; and Earl Smith, former resident of rural Buxton, Albia, Iowa, July 13, 1981.
19. Interviews with Herman Brooks and Hucey Hart.
20. Ibid.
21. *Bystander,* June 28, 1901.
22. Ibid., November 7 and 14, 1902, and April l0, 1908.
23. Ibid., November 14, 1910.
24. Ibid., April 2, 1915.
25. Ibid., March 12, 1909.
26. Ibid., February 6, 1903; January 5, 1906; April 5, May 10, 1907; December 31, 1909; October 30 and December 16, 1910; May 19 and June 23, 1911.
27. Ibid., March 27, 1903, and May 10, 1907.
28. Ibid., February 27, 1903.
29. Ibid., June 9, 1916.
30. Ibid., January 7, 1910, and November 17, 1911.
31. Iowa, Census, 1915, Bluff Creek Township; interviews with Odessa Booker and Mattie Murray.
32. Lewis E. Atherton, *Main Street on the Middle Border* (New York: New York Times Book Company, Quadrangle Books, 1975), pp. 245–49.
33. *Bystander,* July 12, 1907.
34. Ibid., September 6, 1901.
35. Interview with Susie Robinson, former resident of Buxton, Des Moines, Iowa, June 4, 1981.
36. *Bystander,* January 18, 1907.
37. Interviews with Charles Lenger, former resident of Buxton, Oskaloosa, Iowa, July 22, 1981; and Hazel Stapleton, former resident of Buxton, Des Moines, Iowa, June 24, 1980.
38. Interview with Hucey Hart.
39. Interviews with Odessa Booker; and Paul Jackson, former resident of Buxton, Waterloo, Iowa, August 20, 1980.

40. Interviews with Hucey Hart; Lester Beaman; and Odessa Booker.

41. Interview with Jeanette Adams, former resident of Buxton, Des Moines, Iowa, June 21, 1980.

42. Interviews with Wilma Larson Stewart, former resident of rural Buxton, Albia, Iowa, July 21, 1981; and Agnes Erickson, former resident of Buxton, Pershing, Iowa August 15, 1980.

43. Anton Lundeen, *Iowa-konferenses Af Augustana-Synoden Historia* (Augustana Conference, n.d.), p. 1, translation by Karen Cleven, Madrid, Iowa, in possession of Mr. and Mrs. Norman Bergstrom, rural Ogden, Iowa.

44. Ibid., pp. 5, 6.

45. Interview with Alex and Agnes Erickson, former residents of Buxton, Pershing, Iowa, August 15, 1980.

46. Interviews with Wilma Larson Stewart; and Adolph Larson, former resident of Buxton, Knoxville, Iowa, July 23, 1981.

47. Interview with Wilma Larson Stewart.

48. Ibid.

49. Interview with Alex and Agnes Erickson.

50. Interview with Vera Fisher, former resident of Buxton, Newton, Iowa, August 30, 1980.

51. Interviews with Erwin Olsasky, former resident of Buxton, Albia, Iowa, July 15, 1981; Charles Lenger; and Joe Rebarchak, former resident of Buxton, Lovilia, Iowa, July 16, 1981.

52. Interviews with Charles Lenger and Mike Onder.

53. Interview with Erwin Olsasky.

54. Interviews with Emma Romanco Starks, former resident of Buxton, Lovilia, Iowa, July 22, 1981; and Sister Maurine Sofranko, former resident of rural Buxton, Ottumwa, Iowa, September 28, 1980.

55. Interviews with Erwin Olsasky and Charles Lenger.

56. Interview with Emma Romanco Starks.

57. Interview with George Klobnak, former resident of Buxton, Albia, Iowa, June 18, 1981.

58. Ibid.

59. Interviews with Mike Onder and Erwin Olsasky.

60. Interview with Andrew Smith, former resident of Buxton, Lovilia, Iowa, June 17, 1981.

61. Interview with Sister Maurine Sofranko.

62. Interview with Erwin Olsasky.

63. Interview with Emma Romanco Starks.

64. Iowa, Census, 1905 and 1915, Bluff Creek Township.

65. Ibid.

66. Ibid.

67. Ibid.

68. Ibid.

69. Ibid.

70. Interviews with Mike Onder and Earl Smith.

71. Interview with Jacob Brown, former resident of Buxton, Waterloo, Iowa, November 18, 1980.

72. Interview with Harvey Lewis, former resident of Buxton, Bondurant, Iowa, December 29, 1980.

73. Iowa, Census, 1905, Bluff Creek Township.

74. Iowa, Census, 1915, Bluff Creek Township.

75. Sterling D. Spero and Abram L. Harris, *The Black Worker: The Negro and the Labor Movement* (New York: Columbia University Press, 1931), p. 33.

76. Loren Miller, *The Petitioners: The Story of the Supreme Court of the United*

States and the Negro (New York: World Publishing Company, 1967), p. 177; Oliver C. Cox, *Caste, Class and Race* (Garden City, New York: Doubleday and Company, 1970), Chapter 25; and Sig Synnestvedt, *The White Response to Black Emancipation: Second-Class Citizenship in the United States since Reconstruction* (New York: Macmillan Company, 1972), pp. 46–54.

77. Iowa, Census, 1905, Bluff Creek Township.

78. Iowa, Census, 1905 and 1915, Bluff Creek Township.

79. Ibid.; Reports of blacks leaving Buxton appeared in the *Bystander;* see, for example, June 26, 1914, and March 8, 1918.

80. Interview with Jacob Brown.

81. U.S. Department of Commerce, Bureau of the Census, Thirteenth Census of the United States, 1910: Manuscript Population Schedules for Bluff Creek Township, Monroe County, Iowa.

82. Ibid.

83. Ibid.; interview with Odessa Booker.

84. A large number of former residents talked about the high school, including Odessa Booker and Dorothy Collier.

85. Interviews with Robert Wheels, former resident of Buxton, Des Moines, Iowa, June 8, 1981; Helen Clay, former resident of Buxton, Des Moines, Iowa, June 28, 1981; Gertrude Stokes, former resident of Buxton, Cedar Rapids, Iowa, June 15, 1981; and Dorothy Collier.

86. Interviews with Earl Smith and Jacob Brown.

87. Interviews with Alex and Agnes Erickson, Charles Lenger, Wilma Larson Stewart, and Jacob Brown; *Bystander,* September 6, 1912.

88. *Bystander,* May 10, 1912.

89. Interviews with Earl Smith; Vaeletta Fields, former resident of Buxton, Waterloo, Iowa, July 16, 1981; Susie Robinson; Clifford Lewis, former Buxton resident, Eddyville, Iowa, May 28, 1981; and Archie Allison, former resident of Buxton, Albia, Iowa, June 18, 1981.

90. Interviews with Gertrude Stokes; Clara Jones, former resident of Buxton, Lovilia, Iowa, July 22, 1981; and Mike Onder.

91. Interviews with Archie Allison; Gertrude Stokes; and Nellie Lash King, former resident of Buxton, Lovilia, Iowa, December 23, 1980.

92. Interviews with Oliver Burkett, former resident of Buxton, Waterloo, Iowa, November 17, 1980; Naomi Ambey, former resident of Buxton, Cedar Rapids, Iowa, June 16, 1981; Helen Clay; Elmer Buford; and Clyde Wright, former Buxton resident, Waterloo, Iowa, July 20, 1981.

93. Interview with Dorothy Collier.

94. Interviews with Nellie Lash King; Mary Allison, former resident of rural Buxton, Albia, Iowa, February 19, 1981; Dorothy Collier; and Albert and Vera Fisher.

95. Interview with Ada Morgan, former resident of Buxton, Peru, Illinois, July 2, 1981.

96. Ibid.

97. John Hammond Moore, *Albemarle: Jefferson's County, 1727–1976* (Charlottesville: Published for the Albemarle County Historical Society by the University Press of Virginia, 1976), pp. 237–38.

98. Interview with Bessie Lewis.

CHAPTER 7

1. *Iowa State Bystander* (Des Moines), April 26 and September 27, 1912.

2. Ibid., May 3, and November 8, 1912; and May 10, November 16, and Octo-

ber 24, 1913. Even throughout 1911, the number of businesses in Buxton continued to grow. On May 21, 1911 the *Bystander* announced that Buxton had another newspaper, the *Buxton Advocate*. George Lewis opened a "grocery, queensware, glassware and notion Store" in Coopertown in November 1911. The following month, the Monroe Mercantile Company opened a new seventeen-room hotel. In December the main YMCA announced the completion of several remodeling projects. And the next year saw the opening of the Buxton Hotel, owned and operated by Reuben Gaines, Sr. *Bystander,* May 26, November 24 and 28, December 8 and 15, 1911; and September 13, 1912.

3. *Bystander,* January 9, 1914.

4. Ibid., March 27, June 26, July 10, and 24, 1914.

5. Ibid., January 1, 1916; interview with Harold Reasby, former Buxton resident, Waterloo, Iowa, July 21, 1981; Mrs. Henry Brue to Hubert Olin, March 1963, and Ira Wynn to Hubert Olin, March 1963, Olin Papers, Iowa State Historical Department, Des Moines.

6. Interview with Alex Erickson, former Buxton resident, Pershing, Iowa, August 15, 1980.

7. *Albia Republican,* July 24 and April 24, 1913.

8. *Bystander,* November 14, 1913; Olin Papers. Iowa state census data do not support the *Bystander's* claim. According to the census data, Buxton had a white postmaster at least once before 1913.

9. Iowa, *Biennial Report of the State Mine Inspectors, 1915* (Des Moines: State of Iowa, 1915), pp. 28–29; *Biennial Report of the State Mine Inspectors, 1919* (Des Moines: State of Iowa, 1919), p. 46.

10. *The Peabody Atlas* (Chicago: Peabody Coal Company, 1906), p. 44; Iowa, *Biennial Report of the State Mine Inspectors, 1914*, pp. 82–83; and *Biennial Report of the State Mine Inspectors, 1919*, p. 46.

11. *Bystander,* October 10, 1916; June 29, 1917; and February 16, 1917.

12. Ibid., December 29, 1916.

13. Ibid.

14. Iowa, Census of Iowa, 1905 and 1915, Manuscript Population Schedules for Bluff Creek Township, Monroe County.

15. *Bystander,* December 21, 1917.

16. Not a single informant spoke of living in Bucknell, but many people talked of living in Haydock. From this point on, the name Haydock is used to indicate both Haydock and Bucknell. Clayborne Carter to Hubert Olin, March 1963, Olin Papers; and *Bystander,* March 8, 1918, and August 29, 1919.

17. Charles Nichols, *Housing Conditions: Iowa Coal Mining Camps* (Des Moines: State Printing Office, 1919), pp. 20–21.

18. Ibid.

19. Ibid.

20. Reuben Gaines, Jr., "Memoirs of Buxton," 1980, p. 52, in possession of Donald Gaines, rural Albia, Iowa; Clayborne Carter to Hubert Olin, March 1963, Olin Papers; and interview with Archie Harris, former resident of Buxton, rural Lovilia, Iowa, November 15, 1980.

21. "Only Vague Outlines Remain of Thriving Town in 1927," undated *Des Moines Register* clipping in possession of Mary Church, Gillespie, Illinois. Mrs. Church is a former Buxton resident.

22. Interviews with Mike Onder, former resident of Buxton, Des Moines, Iowa, June 23, 1981; Earl Smith, former resident of Buxton, Albia, Iowa, January 1981; and Archie Harris.

23. "Only Vague Outlines Remain of Thriving Town in 1927"; and interview with Lester Beaman, former resident of Buxton, Waterloo, Iowa, November 17, 1980.

24. "Only Vague Outlines Remain of Thriving Town in 1927"; interviews with Earl Smith; Lester Beaman; Mike Onder; Vaeletta Fields former resident of Buxton, Waterloo, Iowa, July 16, 1981; Jacob Brown, former resident of Buxton, Waterloo, Iowa, November 18, 1980; and Mildred Hight Covey, former resident of Haydock, Ames, Iowa, June 14, 1983.

25. Interviews with Lola Reeves, former resident of Buxton, Des Moines, Iowa, June 21, 1980; and Mike Onder; Reuben Gaines, Jr., "Memoirs of Buxton," p. 53.

26. Interviews with Hucey Hart, former resident of Buxton, Des Moines, Iowa, November 1, 1980; Lester Beaman; and Susie Robinson, former resident of Buxton, Des Moines, Iowa, June 4, 1981.

27. Interviews with Hucey Hart; and Nellie Lash King, former resident of Buxton, Lovilia, Iowa, December 23, 1980.

28. Interviews with Nellie Lash King, and Lester Beaman.

29. "Only Vague Outlines Remain of Thriving Town in 1927"; correspondence between F. W. Sargent and Nye Morehouse, assistant general solicitor, Chicago and North Western Railroad, Chicago, Illinois, and stockholders special meeting, November 23, 1925, Consolidation Coal Company, Minutes and Records, vol. 2.

30. Correspondence between F. W. Sargent and Nye Morehouse, Consolidation Coal Company, Records, vol. 2; Reuben Gaines, Jr., "Memoirs of Buxton," p. 54.

31. Iowa, *Biennial Report of the State Mine Inspectors, 1925* (Des Moines: State of Iowa, 1925), p. 85.

32. Ibid., p. 5.

33. Iowa, *Biennial Report of the State Mine Inspectors, 1924* (Des Moines: State of Iowa, 1924), p. 31.

34. Ibid., p. 28.

35. Ibid., p. 32.

36. Iowa, *Biennial Report of the State Mine Inspectors, 1928* (Des Moines: State of Iowa, 1928), p. 3.

37. Interview with Hucey Hart.

38. Interview with Oliver Burkett, former resident of Buxton, Waterloo, Iowa, November 17, 1980.

39. Reuben Gaines, Jr., "Memoirs of Buxton," p. 53.

40. Ibid., p. 54.

41. Interviews with Carl Kietzman, former resident of Buxton, rural Albia, Iowa, May 28, 1981; and Mike Onder; Reuben Gaines, Jr., "Memoirs of Buxton," p. 53.

42. "Only Vague Outlines Remain of Thriving Town in 1927."

43. Vaeletta Fields to Hubert Olin, March 1963, Olin Papers.

44. The families included in this portion of the study were selected because of the completeness of the information in the oral histories. In some cases, further information was obtained by correspondence.

45. Interview with Vaeletta Fields.

46. Interviews with Alex Erickson; and Catherine McClelland, former Zookspur resident Madrid, November 1972; and William Ewalt, "The Ogden Coal Strike, 1910–1912," (Student paper, Iowa State University, 1974), pp. 5–10, in possession of Dorothy Schwieder.

47. Interviews with Gertrude Stokes, former resident of Buxton, Cedar Rapids, Iowa, June 15, 1981; and Hucey Hart.

48. Interview with Lara Wardelin, former resident of Buxton, Des Moines, Iowa, July 8, 1981.

49. Interview with Odessa Booker, former resident of Buxton, Des Moines, Iowa, June 8, 1981.

50. Interviews with Alex and Agnes Erickson, former residents of Buxton, Pershing, Iowa, August 15, 1980.

51. Interview with Mattie Murray, former resident of Buxton, Des Moines, June 4, 1981.

52. Interview with Marjorie Brown, former resident of Buxton, Waterloo, Iowa, August 20, 1980.

53. Ibid.

54. Ibid.

55. Marian Carter to Hubert Olin, October 1962 and January 1963, Olin Papers.

56. *Bystander,* August 29, 1919; and interview with Vaeletta Fields.

57. Interviews with Reuben Gaines, Jr., former resident of Buxton and Haydock, rural Albia, Iowa, February, 19, 1981; and Archie Harris.

CHAPTER 8

1. Hubert Olin Papers, Iowa State Historical Department, Des Moines, Iowa.

2. Interviews with Archie Harris, former resident of Buxton, Lovilia, Iowa, February 19, 1981; Hucey Hart, former resident of Buxton and Haydock, Des Moines, Iowa, November 1, 1980; and Robert Wheels, former resident of Buxton and Haydock, Des Moines, Iowa, June 8, 1981.

3. Olin Papers; and interview with Ada Morgan, former resident of Buxton, Peru, Illinois, July 2, 1981.

4. A. E. Erskin, to Hubert Olin, April 1963, Olin Papers.

5. David Corbin, *Life, Work, and Rebellion in the Coal Fields: The Southern West Virginia Miners, 1880-1922,* The Working Class in American History Series (Urbana: University of Illinois Press, 1981), pp. 65-70.

6. Interview with Hucey Hart.

7. Interview with Marjorie Brown, former resident of Buxton, Waterloo, Iowa, August 20, 1980.

SELECTED BIBLIOGRAPHY

GOVERNMENT DOCUMENTS

Iowa. *Biennial Report of the State Mine Inspectors,* 1880–1925. Des Moines: State of Iowa.

Iowa. Census of Iowa, 1885 and 1895. Manuscript Population Schedules for Harrison and East Des Moines Townships, Mahaska County. Iowa State Historical Department, Des Moines.

Iowa. Census of Iowa, 1905 and 1915. Manuscript Population Schedules for Bluff Creek Township, Monroe County; and for Beacon, Cincinnati, and Seymour, Iowa. Iowa State Historical Department, Des Moines.

Iowa Official Register, 1915–16.

Laws of Iowa, 1880–1925. Des Moines: State Printer, 1880–1925.

Monroe County, Iowa. Bluff Creek Township Birth Records, 1904–1905. Monroe County Sheriff's Office, Albia, Iowa.

U.S. Census Office. *Sixth Census of the United States, 1840.*

U.S. Department of Commerce, Bureau of the Census. *Eleventh Census of the United States, 1890: Population,* vol. 2, pt. 1. Washington, D.C.: Government Printing Office, 1895.

U.S. Department of Commerce, Bureau of the Census. Twelfth Census of the United States, 1900: Manuscript Population Schedules for Harrison and East Des Moines Townships, Mahaska County, Iowa.

U.S. Department of Commerce, Bureau of the Census. Thirteenth Census of the United States, 1910: Manuscript Population Schedules for Bluff Creek Township, Monroe County, Iowa.

NEWSPAPERS AND PERIODICALS

Albia Republican, 1910–1915.

The Bystander (Des Moines), 1916–1925.

Eddyville Tribune, 1898.

Iowa State Bystander (Des Moines), 1896–1916.

Muchakinock State, 1898.

The Negro Solicitor (Oskaloosa), 1898.

Oskaloosa Daily Herald, 1940.

Oskaloosa Daily News, 1897.

Oskaloosa Weekly Herald, 1883.

United Mine Workers Journal, 1891–1925.

MANUSCRIPTS

Consolidation Coal Company. Minutes of business meetings and records. 2 vols. Chicago and North Western Railroad, Chicago, Illinois.

Cornelius, Jane. "The Blacks in Waterloo." Student paper, Iowa State University, 1972. In possession of Dorothy Schwieder.

Ewalt, William. "The Ogden Coal Strike, 1910–1912." Student paper, Iowa State University, 1974. In possession of Dorothy Schwieder.

Gaines, Reuben, Jr. "Memoirs of Buxton." 1980. In possession of Donald Gaines, rural Albia, Iowa.

London, Minnie B. "As I Remember." 1940. Iowa State Historical Department, Des Moines.

Olin, Hubert L., Papers. Iowa State Historical Department, Des Moines.

United Mine Workers of America. District 13. Executive Board Proceedings, 1891–1925. Microfilm, 4 rolls. Iowa State Historical Department, Des Moines.

United Mine Workers of America. District 13. 1910–1912 Joint Wage Agreement. Iowa State Historical Department, Des Moines.

BOOKS

Athearn, Robert G. *In Search of Canaan: Black Migration to Kansas, 1879–80.* Lawrence: Regents Press of Kansas, 1978.

Atherton, Lewis E. *Main Street on the Middle Border.* New York: New York Times Book Company, Quadrangle Books, 1975.

Bergmann, Leola Nelson. *The Negro in Iowa.* Iowa City: State Historical Society of Iowa, 1969.

Bodnar, John, Roger Simon, and Michael P. Weber. *Lives of Their Own: Blacks, Italians, and Poles in Pittsburgh, 1900–1960.* The Working Class in American History series. Urbana: University of Illinois Press, 1982.

Brandes, Stuart D. *American Welfare Capitalism, 1880–1940.* Chicago: University of Chicago Press, 1976.

Broom, Leonard, and Norval Glenn. *Transformation of the Negro American.* New York: Harper and Row, 1965.

Byington, Margaret F. *Homestead: The Households of a Mill Town.* The Pittsburgh Survey, Vol. 4. New York: Russell Sage Foundation, 1910.

Corbin, David. *Life, Work, and Rebellion in the Coal Fields: The Southern West Virginia Miners, 1880–1922.* Urbana: University of Illinois Press, 1981.

Cox, Oliver C. *Caste, Class and Race.* Garden City, N.Y.: Doubleday and Company, 1970.

Ginger, Ray. *Altgeld's America: The Lincoln Ideal Versus Changing Realities.* New York: Funk and Wagnalls Company, 1958.

Goodrich, Carter. *The Miner's Freedom: A Study of the Working Life in a Changing Society.* Boston: Marshall Jones Company, 1925.

Gutman, Herbert G. *The Black Family in Slavery and Freedom, 1750–1925.* New York: Vintage Books, 1977.

A History of Lucas County. Des Moines: State Historical Company, 1881.

The History of Mahaska County. Des Moines: Union Historical Company, 1878.

Hraba, Joseph. *American Ethnicity*. Itasca, Ill.: F. E. Peacock, Publishers, 1979.

Lundeen, Anton. *Iowa-Konferenses af Augustana-Synoden Historia*. Augustana Conference, n.d. Translation by Karen Cleven, Madrid, Iowa. In possession of Mr. and Mrs. Norman Bergstrom, rural Ogden, Iowa.

Miller, Loren. *The Petitioners: The Story of the Supreme Court of the United States and the Negro*. New York: World Publishing Company, 1967.

Moore, John Hammond. *Albemarle: Jefferson's County, 1727–1976*. Charlottesville: Published for the Albemarle County Historical Society by the University Press of Virginia, 1976.

Nichols, Charles. *Housing Conditions: Iowa Coal Mining Camps*. Des Moines: State Printing Office, 1919.

Norton, Mary Beth, et al. *A People and a Nation: A History of the United States*. Vol. 2. Boston: Houghton Mifflin Company, 1982.

Olin, Hubert L. *Coal Mining in Iowa*. Des Moines: State of Iowa, 1965.

The Peabody Atlas. Chicago: Peabody Coal Company, 1906.

Polk, R. L., and Company. *Iowa State Gazetteer and Business Directory*. 1882–1883, 1884–1885, 1906. Des Moines: R. L. Polk and Company, 1882, 1884, 1906.

Sage, Leland L. *A History of Iowa*. Ames: Iowa State University Press, 1974.

Schneider, Eugene V. *Industrial Sociology: The Social Relations of Industry and the Community*. 2d ed. New York: McGraw-Hill Book Company, 1969.

Schwieder, Dorothy. *Black Diamonds: Life and Work in Iowa's Coal Mining Communities, 1895–1925*. Ames: Iowa State University Press, 1983.

Shannon, Fred A. *The Farmer's Last Frontier: Agriculture, 1860–1897*. Vol. 5 of The Economic History of the United States, ed. Henry David. New York: Farrar and Rinehart, 1945.

Shurick, A. T. *The Coal Industry*. Boston: Little, Brown and Company, 1924.

Spero, Sterling D., and Abram L. Harris. *The Black Worker: The Negro and the Labor Movement*. New York: Columbia University Press, 1931.

Synnestvedt, Sig. *The White Response to Black Emancipation: Second-Class Citizenship in the United States since Reconstruction*. New York: Macmillan Company, 1972.

Wall, Joseph Frazier. *Iowa: A Bicentennial History*. New York: W. W. Norton and Company; Nashville: American Association for State and Local History, 1978.

Woodward, C. Vann. *The Strange Career of Jim Crow*. New York: Oxford University Press, 1955.

ARTICLES

Baker, Donald. "Operating Conditions Encountered in Iowa." *Coal Age* 14 (March 1921): 437–41.

Hill, James L. "Migration of Blacks to Iowa, 1820–1960." *Journal of Negro History* 66 (Winter 1981–82): 289–303.
Keyes, Charles. "Historical Sketch of Mining." *Iowa Geological Survey* 22 (1912): 89–121.
Lees, James H. "History of Coal Mining in Iowa." *Iowa Geological Survey* 19 (1908): 525–88.
McLaughlin, David. "Mahaska County Once Led State in Coal Mining Activities." Reprint from Work Projects Administration report, Iowa Writers' Program, *Oskaloosa Daily Herald*, January 30, 1940.
Pleck, Elizabeth. "A Mother's Wages: Income Earning Among Married Italian and Black Women, 1896–1911." In *A Heritage of Her Own: Toward a New Social History of American Women*, ed. Nancy F. Cott and Elizabeth Pleck, pp. 367–92. New York: Simon and Schuster, 1979.
Rutland, Robert. "The Mining Camps of Iowa: Faded Sources of Hawkeye History." *Iowa Journal of History* 54 (January 1956), 35–42.
Rye, Stephen H. "Buxton: Black Metropolis of Iowa." *Annals of Iowa* 41 (Spring 1972): 939–57.
Sheridan, Frank J. "Italian, Slavic and Hungarian Unskilled Immigrant Laborers in the United States." *Labor Bureau Bulletin* 15 (1907): 403–86.
Shiffer, Beverly. "The Story of Buxton." *Annals of Iowa* 37 (Summer 1964): 339–47.
Swisher, Jacob A. "Mining in Iowa." *Iowa Journal of History and Politics* 43 (October 1945): 305–56.
______. "The Rise and Fall of Buxotn." *Palimpsest* 26 (June 1945): 179–92.

INTERVIEWS

Adams, Jeanette. Des Moines, Iowa, June 21, 1980.
Allison, Archie. Albia, Iowa, June 18, 1981.
Allison, Mary. Albia, Iowa, February 19, 1981.
Ambey, Naomi. Cedar Rapids, Iowa, June 16, 1981.
Antolik, Andy. Albia, Iowa, June 18, 1981.
Aubrey, Dean. Ames, Iowa, August 2, 1972.
Barnes, Mildred. Lovilia, Iowa, June 12, 1981.
Beaman, Lester. Waterloo, Iowa, November 17, 1980.
Blomgren, Emmanuel. Rural Albia, Iowa, July 15, 1981.
Booker, Odessa. Des Moines, Iowa, June 8, 1981.
Bragg, Nina. Cedar Rapids, Iowa, June 15, 1981.
Brooks, Herman. Chicago, Illinois, July 30, 1981.
Brown, Jacob. Waterloo, Iowa, November 18, 1980.
Brown, Marjorie. Waterloo, Iowa, October 15, 1978, and August 19 and 20, 1980.
Buford, Elmer. Des Moines, Iowa, June 30, 1980.
Burkett, Oliver. Waterloo, Iowa, November 17, 1980.
Clay, Helen. Des Moines, Iowa, June 28, 1981.
Collier, Dorothy. Des Moines, Iowa, November 14, 1978, and October 17, 1980.

Covey, Mildred Hight. Ames, Iowa, June 14, 1983.
Cox, Albert Harvey. Newton, Iowa, July 31, 1980.
Duke, Helen. Des Moines, Iowa, June 30, 1981.
Erickson, Agnes. Pershing, Iowa, August 15, 1980.
Erickson, Alex. Pershing, Iowa, August 15, 1980.
Fields, Vaeletta. Waterloo, Iowa, July 7 and 16, 1981.
Fisher, Albert. Newton, Iowa, August 30, 1980.
Fisher, Vera. Newton, Iowa, August 30, 1980.
Franzen, Marvin. Lovilia, Iowa, September 9, 1981.
Frazier, Jessie. Des Moines, Iowa, March 15, 1979.
Gaines, Donald. Rural Albia, Iowa, June 12, 1981.
Gaines, Reuben, Jr. Rural Albia, Iowa, February 19, 1981.
Gardner, Mildred. Albia, Iowa, June 4, 1981.
Gardner, Walter. Albia, Iowa, June 4, 1981.
Goodwin, Carl. Knoxville, Iowa, December 23, 1980.
Goodwin, Irene. Lovilia, Iowa, June 28, 1983.
Haluska, Helen Lesko. Albia, Iowa, June 17, 1981.
Harris, Archie. Rural Lovilia, Iowa, November 15, 1980, and February 19, 1981.
Hart, Hucey. Des Moines, Iowa, November 1, 1980.
Holm, Gus. Ames, Iowa, June 20, 1980.
Jackson, Mary. Waterloo, Iowa, August 20, 1980.
Jackson, Paul. Waterloo, Iowa, August 20, 1980.
Jones, Clara. Lovilia, Iowa, July 22, 1981.
Kietzman, Carl. Rural Albia, Iowa, May 28, 1981.
King, Nellie Lash. Lovilia, Iowa, December 23, 1980.
Klobnak, George. Albia, Iowa, June 18, 1981.
Larson, Adolph. Knoxville, Iowa, July 23, 1981.
Lenger, Charles. Oskaloosa, Iowa, July 22, 1981.
Lewis, Bessie. Des Moines, Iowa, January 17, 1981.
Lewis, Clifford. Eddyville, Iowa, May 28, 1981.
Lewis, Harvey. Bondurant, Iowa, December 29, 1980.
McClelland, Catherine. Madrid, Iowa, November 4, 1972.
Mellick, Joe. Albia, Iowa, June 16, 1981.
Morgan, Ada. Peru, Illinois, July 2, 1981.
Murray, Mattie. Des Moines, Iowa, June 4, 1981.
Nizzi, Lola. Granger, Iowa, January 10, 1978.
Olsasky, Erwin. Albia, Iowa, July 15, 1981.
Onder, Mike. Des Moines, Iowa, June 23, 1981.
Onder, Steven. Lovilia, Iowa, July 14, 1981.
Papich, Milo. Slater, Iowa, May 8, 1979.
Perot, John. Albia, Iowa, June 19, 1981.
Plum, Walter. Knoxville, Iowa, July 22, 1981.
Reasby, Harold. Waterloo, Iowa, July 21, 1981.
Rebarchak, Joe. Albia, Iowa, July 16, 1981.
Reeves, Lola. Des Moines, Iowa, June 21, 1980.
Robinson, Susie. Des Moines, Iowa, June 4, 1981.
Slofkosky, Steve. Albia, Iowa, June 18, 1981.

Smith, Andrew. Lovilia, Iowa, June 17, 1981.
Smith, Earl. Albia, Iowa, January 4 and July 13, 1981.
Sofranko, Sister Maurine. Ottumwa, Iowa, September 28, 1980.
Sofranko, Eleanor. Lovilia, Iowa, July 19, 1981.
Stapleton, Hazel. Des Moines, Iowa, June 24, 1980.
Starks, Emma Romanco. Lovilia, Iowa, July 22, 1981.
Stein, Maurice. Ottumwa, Iowa, July 29, 1981.
Stewart, Wilma Larson. Albia, Iowa, July 21, 1981.
Stokes, Gertrude. Cedar Rapids, Iowa, June 15, 1981.
Taylor, Charles. Des Moines, Iowa, July 8, 1980.
Walker, Agnes. Cedar Rapids, Iowa, June 15, 1981.
Walraven, Lou. Des Moines, Iowa, January 31, 1982.
Wardelin, Lara. Des Moines, Iowa, July 8, 1981.
Wheels, Robert. Des Moines, Iowa, June 8, 1981.
Wright, Clyde. Waterloo, Iowa, July 20, 1981.
Wright, Leroy. Waterloo, Iowa, November 18, 1980.

INDEX